D1491601

She, then, like snow in a dark night,
Fell secretly. And the world waked
With dazzling of the drowsy eye,
So that some muttered 'Too much light',
And drew the curtains close.
Like snow, warmer than fingers feared,
And to soil friendly;
Holding the histories of the night
In yet unmelted tracks.

ROBERT GRAVES

Published by Little Toller Books in 2016
Little Toller Books, Lower Dairy, Toller Fratrum, Dorset

Words © Marcus Sedgwick 2016

The right of Marcus Sedgwick to be identified as the author of this work has
been asserted in accordance with Copyright, Design and Patents Act 1988

Photographs by Wilson Bentley courtesy of the Smithsonian Institution Archives

Typeset in Garamond and Perpetua by Little Toller Books

Printed by TJ International, Padstow, Cornwall

All papers used by Little Toller Books are natural, recyclable products made
from wood grown in sustainable, well-managed forests

A catalogue record for this book is available from the British Library

All rights reserved

ISBN 978-1-908213-40-2

02

Acknowledgements

Many thanks to the Little Toller subscribers who have supported this
publication: Robert Goddard, Nima Reid, Claire and John Plass, Harriet
Owens, Neil Confrey, Tanya Bruce-Lockhart, Janice Stockford, Joe Wright,
Rosemary Fern, Michael Hunt, Ian McMillan, Melly Nile and Hugo Rhys,
Naomi Nile, Georgina Tuson, Bob Buhr, Christine Shaw, Alexa de Ferranti,
Jonathan Clarke, Ron Riley, Keith Halfacree, Howard Wix, Mattijs Baarsma,
Richard Brett, Richard Brown, Barry Taylor, Gareth Watson, Paddy Bullard,
Elizabeth Shaw, Rosalina Pounder, Graham Powell, Timothy Boswell, Peter
Reynolds, David Broom, Penny Hodgkinson, Suzanne Stock, Inigo Roque
Egusquiza, Leonie Hicks and Jo Sweeting. Thanks also to the UK Met Office for
their assistance during the research of this book.

Snow

MARCUS SEDGWICK

A LITTLE TOLLER **MONOGRAPH**

Contents

À Yves, qui n'aime pas la neige.

Snow Talk

Yesterday saw the first snowfall of the year. A light dusting in the morning turned out to be merely the prelude to a thick, wet and heavy fall later in the afternoon. Today, it's vanished again, as fast as it came – a warm bathing of sun was enough to melt it all. There have been bright, warm days of late; an Indian Summer with no cloud in sight. Looking down the valley now, between the humps of the nearby hills that lie like a line of vast migrating mammoths, the tip of the summit of Mont Blanc is making one of its rare appearances *sans chapeau – without a hat –* above the cloud that almost perpetually envelops the peak. But yesterday the snow laid down its marker for the season, made its opening move, letting us know it's on its way in earnest. It is October 16.

We live in the Haute-Savoie, an alpine department of eastern France bordering both Switzerland and Italy. Our house sits in a liminal space, halfway from the town below to the high plain above, where the resort village nestles underneath a mountain that is 2,000 metres high. That puts our house at an altitude of 1,100 metres, something that pleased me to discover after we'd moved in, this being more or less the height of the summit of Snowdon, a mountain on which I spent considerable time as a child, during summer holidays.

All my life I have loved the snow; ever since boyhood, when it seemed that every year was blessed, if you see it that way, with a heavy snowfall. There were hard winters when I was a child, growing up in the Far East. The far east that is, of Kent, close enough to the continent to be swept by their snow clouds, and it was an utter delight for my brother and me when our parents declared that the roads from our village to school, only a few miles away, were impassable once again. It didn't take much snow to do it; the sunken rural roads that ran out of the orchards and across open wind-swept cabbage fields were easily prone to collect drifting snow, sealing us happily into our tiny village.

Imprisoned, we were, ironically, given something wonderful: freedom! No school, for maybe two or three days at a time; time that had been unexpectedly given back to us by the snowfall. It's hard not to link some images; there, in the house, on the bookshelves, our father's collection of books by William Blake. Lurking inside which somewhere would have been Blake's verses depicting *The Schoolboy*, trudging as miserably towards education as he would towards jail. There too would have been Blake's lines *Soft Snow*:

> *I walked abroad in a snowy day;*
> *I asked the soft snow with me to play.*

And play we did. My brother and I pulled on woollen scarves and gloves; in those days before miracle fabrics our hands were probably colder from snow-sodden wool than they would have been bare. Nevertheless, despite stinging red hands, we were free to roam in our large garden, still wild and overgrown in many places despite our parents' efforts to tame it. Snowmen, snowball fights, icicles as daggers or as ice-dragon teeth, the snow seems to me now to have been a (literally) brilliant canvas for the imagination. Tiring, we would stand at the garden gate, maybe dimly aware of the strangeness, the silence of the day; with no cars passing and just a few traces of footsteps in the otherwise untouched snow. Then we'd head out across the shallow valley below the house, over snow-bound fields, into the beech copse, or down to the slow-moving, meandering river, where we would dare each other to test the ice.

Is this why I have a life-long devotion to the stuff, simply because of the freedom it represented as a child; salvation from school, which both my brother and I hated and feared? Or is it something deeper than that? Either way, as I became a writer and I began producing novels, I became aware that I often returned to snow in my fiction, deliberately choosing locations for stories that were frozen and white.

It's something I've come to think about more seriously, now that we live full time in the mountains. From our first full winter here, two conversations stick out of the memory landscape; first, the owner of a hotel further up the valley, who asked us if we had ever cleared snow. Yes, of course, we said. I cleared plenty of it as a child, in Kent, and then in Wales when my family moved there. Maureen, my wife, grew up on Long Island, and New York can freeze in the winter just as hard as it bakes in the summer. Yes, we said, we've cleared snow. The hotel owner looked at us. No, he said. Have you really cleared Snow? This time, I could see the capital S on the word.

That winter, we found out what he meant. You awake and find half a metre has fallen, or more, overnight. You wrestle into clothes and boots and set to; clearing the path from the house, twenty metres up to the road, from the path down the steps to the terrace, across the terrace to the woodshed, across the other way to the tool store. Done. Two hours later, just as you recover, you look out and see another thirty centimetres has fallen. You do it all again. And then, a little later in the afternoon, another thirty centimetres has fallen, and you do it all for a third time. Of course, days like that are unusual, but once you've experienced one, you are given a new respect for snow.

Clearing snow is one of those things that is more complicated than it first appears. The first thing: you learn not to walk on freshly fallen snow – packing it down makes it icy, it becomes both tricky to walk on and hard to shift. Equally, at either end of the winter, just the wrong amount of rain can turn the light, easily shovelled fluff into heavy slush, and, if temperatures drop again, it freezes into ice that would make very healthy blocks for igloos. One can buy a *fraise-à-*

neige of course, a mechanical snow blower, to chomp and spit its way through the drifts, but even then there are things to consider, the main one being, where am I blowing my snow *to*? The snow can arrive in October. It can still be falling in May. The issue of where you move these ever growing piles of snow means that you can very easily run out of room, and wall yourself in. That first winter was a hard one; by the end of it we were walking to the front door through a tunnel higher than I am tall. These days we are careful to keep several paths clear to the edge of the terrace – we live on a 45 degree mountainside – and let gravity take the snow away for us.

The second conversation was even simpler. Buying some rough planks from the sawmill halfway down to flatland, Yves, the gently-spoken *scieur,* looked up at the sky. It was October, and there were ominous clouds. You bought a house in the mountains? Yes, we said. You like snow? Yes, we said. He blinked a couple of times, then frowned, deadly serious. I hope you do, he said.

In order to understand a thing, I have always felt it important to know a little about its origins. This is just as true for words as for things. I don't think it's possible to be a writer and not be at the very least occasionally interested in their etymology. Some words more than others are self-contained history lessons; witness our words for livestock: we eat 'pork' and 'beef', but the animals we rear are called 'pigs' and 'cows', the difference being that the names for the meat we eat were given to us by our Norman invaders and rulers, while the livestock was tended for them by the now subjugated population, speaking tongues derived from Old English.

Are such things important? I think they are, but then I'm a writer. I do however believe that even if the reader is not consciously aware of it, the mood of a piece of writing is powerfully, even though perhaps subtly, changed by such choices. Part of the reason why English is such a rich and fascinating language (not to mention one with so many irregularities and contradictions) is that it is a historical melding of different language groups; modern English is a mongrel, not of ill-breeding but, I would argue, of spectacularly *good* breeding. This is not the place to go into a history of our language but what we have ended up with today is a mixture of Germanic languages (Old and Middle English, Old Norse, Dutch), French, Latin and Greek along with a smattering of obscure, and even a few unknown, sources. In any given sentence therefore, the writer of English is faced with a multitude of synonyms for the concept he or she wishes to express. *Eructate* and *belch* both mean the same thing; one derives from Latin and one from Old English, but they are massively different in terms of feeling and their usage depends highly on that fact and the context in which they are to be deployed. So too with many other words in English; there will always be a choice between several synonyms, but each has a subtly different feel, and subtlety is the hidden heart of writing; it's what overturns cliché and transforms it into virtuosity.

So what about a word like snow, does it have a history to tell us? It's immediately obvious that words like this, which are as fundamental to life in Northern Europe as any, are not only very old, but very likely share a common ancestor. This is indeed the case. Our modern word *snow* derives from what linguists call Proto-Indo-European: a surmised common ancestor of all Indo-European languages. The word given in PIE is *sniegwh,* which in turn moves into Proto-Germanic as *snaiwaz,* and which we can recognise in modern German *Schnee*, Dutch *sneeuw*, Irish *sneachd,* Slovak *sneh,* Swedish *snö*, and so on, as well as our own word. In French *(neige)* and Spanish *(nieve),* the initial S was lost in the shift from PIE to Latin and then the Romance branches, but the word is essentially the same. This being an Indo-European family, we can even trace its survival into Sanskrit *(snihyati)* though its meaning has shifted over time to mean 'gets wet'. And this is all as we would expect, since snow has been a fundamental part of life in many parts of the world from as soon as we learned to use language (and indeed before). Its history tells us so.

This raises an immediate thought. Is the opposite therefore the case; that parts of the world that have never seen a snowflake do not have a word for snow in their language? To answer this we step foot into the hazardous world of linguistics, where certainty is very hard to come by. Spend a little time reading papers by linguists and you will soon discover their contempt for lazily repeated and often unverifiable claims, such as suggestions of the kind that there is no word for snow in Arabic. (There is, it's ثلج and although this word can also mean ice, as with any language, context is usually sufficient to make it clear what is meant.) What seems to really bug linguists, however, is that these claims don't die easily, despite perpetual attempts to put them to rest. Remember Reagan's claim in the 1980s that there is no word for 'freedom' in Russian? It seems too ludicrous to be true, but these words were actually uttered in an interview with the BBC in October 1985. A statement of this order might seem preposterous to us today, but the political climate of those days meant there were all too many people wanting to believe such a thing, and wanting is more than half the battle. Language is the tool of the propagandist; it has been for a very long time and always will be.

Why any of this matters is all to do with a thing called the Sapir-Whorf hypothesis. Edward Sapir was an anthropological linguist, and teacher of Benjamin Lee Whorf. Working in the early part of the twentieth-century, their hypothesis was that the words available for use in any given language significantly determine the thought processes of the speakers of that language, and thus that speakers of different languages think about the world in very different ways. (No word for freedom? That should keep the serfs from revolting.) It's a theory that has come under frequent attack, and while some of its more extravagant claims have been dissected and dropped, the general idea lives on under the name of linguistic relativism. It's a powerful idea. Of course, it's immediately obvious that the Ancient Greeks didn't have a word for 'television' and that therefore the Greeks spent no time thinking about televisions. (Ironically that word is half Greek and half Latin, a fact which led C.P. Scott, legendary editor of the *Manchester Guardian*, to declare 'no good will come of this device'). Such truisms get us nowhere, but are there more thought-provoking examples of how language might shape thought?

The answer to that has linguists in huddles in conferences, sending their champions out to fight their corner from time to time. It is, however, apparent that languages work differently. Having seen my books translated into many languages, some, like Swedish and German, closer to English, and some very, very different, such as Japanese and Russian, and having had many conversations over the years about the intricate problems of translation, it is obvious that some things are easier to translate than others. These problems are what make translation such an art form; if it were a simple matter of word substitution, Google Translate would be a hell of a lot more accurate. As related as German and English are, the simple fact that German sends its verb infinitives to the end of a sentence can drastically alter the way the tension of a sentence is deployed when shifted into English. Can a translator match such things? I'm reminded of the quote that the translator Daniel Hahn likes to use when asked about the difficulties of translation, concerning Ginger Rogers. When asked how amazing it was to have done all those dances with Fred Astaire, she supposedly replied, 'I did everything he did, backwards and in high heels.'

As problematic as translation can be, that is not to say that a certain idea cannot be expressed at all in a different language, and it is a very long way from the notion that a thing cannot be *thought about*. This is what drives linguists crazy; because it is always possible in any language to 'find' words that 'do not exist', but that does not mean that the concept does not exist, or cannot be expressed. There is no word in English for the growing sensation that you are listening to a stupidly short loop of *Muzak* while dining in an otherwise fine restaurant, but I've just expressed the concept, albeit pretty clumsily.

So even if it were the case that a people who had never seen snow had no word for it, they would soon do what any language does when faced with such an issue, namely make one up, or borrow it from another language. To see the truth of that you only need to consider 'English' words (by which I mean ones to be found in any standard English dictionary) such as déjà-vu, ennui, kindergarten or jungle.

Without question, however, the single issue in linguistics that finds its way into popular culture on the most regular basis is the assertion that the Inuit have fifty words for snow. Or a hundred. Or a thousand. So prevalent is this concept that it has spawned a neologism: *snowclone*, a word that expresses the concept of a phrasal template that becomes so popular, familiar or clichéd that it can be adapted to new situations and concepts at the whim of the speaker. An example of this would be the phrase reputedly born from the fashion industry, 'grey is the new black.' So recognisable is this construction that we can now say 'Thursday is the new Friday', 'forty is the new thirty' or whatever combination we like and know that we will be understood, and may well be understood to be referencing the original saying. Or not. But it all began with those Inuit, and the idea that 'if the Inuit have fifty words for snow, then the X must have fifty words for Y.'

So what about the Inuit? Do they really have copious words for snow? Much as with the easy desire for many in the West to swallow Reagan's statement about Russia and freedom, it's an idea that we *want* to be true. The Inuit live in a vast white wilderness, 99 per cent of everything they see and touch must surely be snow or ice; no wonder they have multiple words for the stuff in all its forms; variations of such subtle complexity that we pampered rich folk could never understand them. So the argument runs, in a split second, in our thoughts, and the statement is accepted.

The idea was born around a hundred years ago, when anthropologist Franz Boaz described his findings about the peoples of the North in his *Handbook of American Indian Languages*, published in 1911. In the introduction he notes how the Eskimo (the term in use at the time) had different words for falling snow, snow on the ground, drifting snow, and so on and so on. Boaz did some good things in his career: he was outspoken against the then current trends of scientific racism and white superiority, for example. But he left behind an idea that, after lying dormant like a forgotten landmine for decades, exploded into an argument that still rages. As recently as 2013, the *Washington Post* ran a piece declaring 'There really are fifty Eskimo words for snow', which most people would have read and thought *There, I knew it had to be true* but which led certain linguists to froth at the mouth.

One such linguist, Laura Martin, re-evaluated Boaz's claims in the 1980s. In what would later be called the Great Eskimo Vocabulary Hoax, others backed up her criticisms, and yet the belief refuses to die, fuelled in no small part by other linguists, such as those quoted in the Washington Post article, who sought to prove that Inuit does have multiple words for snow.

Much of the argument has been driven by differences in interpretation of what the apparently simple question *How many words do the Inuit have for snow?* actually means. What do we mean by Inuit for example? And more importantly, what do we mean by *word?* Do we count every individual word, or the basic word stem or root, what a linguist would call a lexeme, and not all the possible derivations and inflections of that lexeme?

Ultimately, the arguments are pointless. This sweeping statement can be justified by noting how Inuit languages work. What was once blanket termed 'Eskimo' is actually composed of two distinct language groups, Inuit and Yupik, each with many distinct languages of their own. The critical point is that Inuit is a polysynthetic language, which means that new words can be formed by compounding existing ones. So Geoffrey Pullum, amongst others, has argued that while there are a few distinct words for snow in the Inuit languages, words meaning snow on the ground, slush, drift, falling snow, etc., much as Boaz said, that due to the polysynthetic nature of these languages, one could go on forever making new compounds for any particular circumstance one came across. Thus he argues there are, in theory at least, an infinite number of words for snow in Inuit. The truth, Pullum concludes, is that the number of words is bounded only by one's stamina.

So we don't really know if it's true that the Inuit have '50 words for snow'. Kate Bush may or may not have consulted an anthropologist before writing a song and album of that name, but she explains that it was her fascination with the fifty-word myth that led her to begin to create her own, mythological words for snow. And that is absolutely an artist's prerogative, or, in fact, *need*; to create from fantasy as much as from truth. The idea that the Inuit have 50 words for snow is an idea we want to be true, it will probably never die, and it expresses a truth, even if it itself is false.

It's hard not to feel that the debate around the Great Eskimo Language Hoax could have been curtailed if attention had been paid to a different language, one that is not polysynthetic, but which exists in a similarly snowy clime. Sami is such a language group, and studies of the indigenous peoples in the far North of Norway, Sweden and Finland have shown us that perhaps our first instincts were not as untrue as we've been told. These recent studies have shown that Sami languages have perhaps more than 180 *distinct* snow and ice-related words, and maybe more than 300 if we include words for tracks in snow, snow conditions and so on.

So what about English? While it cannot contend with the Sami languages' depth, we do have a few distinct words of our own. A few moments thought would produce a list that includes snow, sleet, hail, slush, flake, and then we might add related terms: ice, icicle, snowdrift and blizzard. We could add some terms familiar to the skier: pack, crust, powder. And we could also add in some less familiar and borrowed words from other languages, but which nevertheless reside in the English dictionary, words like *graupel, firn* and *névé*.

Graupel, a German word, describes snow that falls as tiny pellet-like entities that form when super-cooled droplets of water freeze around a falling snowflake. Such a thing is somehow extremely beautiful to even think about. Even here we must be careful with our terminology, for a snow *pellet* is itself a distinct thing according to meteorologists. Pellets look like small hail but form in winter, unlike hail, which is born of thunderstorms.

Névé is French, and supposedly derives from the area where we live; the Savoyard dialect word *nevi* means a mass of snow. *Névé* is also granular snow, which has fallen, partially melted and refrozen. If it survives an entire season it will become known as *firn*; older and denser, it is the beginning of the formation of a glacier, and in this word we have another piece of etymological archaeology – for *firn* is the last surviving relic in English of a more general Indo-European word that meant 'of last year'. German still holds the term *Firnewein* – last year's wine – for example. And Old English had *fyrngemynd*, meaning 'ancient history', but which could more literally be translated as 'the memory of long ago'. Which, I think, is exactly what is brought to mind when gazing upon a dwindling but tenacious patch of old snow on a summer mountainside.

Enough. For now, it's sufficient to notice again that when the need arises, language, like a living thing, creates or borrows the words it needs, and these words themselves have beauty and power, which is what drives poets and writers to consider their subtleties and complexities and then deploy them to best effect.

Snow! Just the word alone, as a single-word statement, is enough to conjure a powerful reaction in many of us. This was well noted by Llewellyn Powys, member of that extraordinarily talented family of eleven siblings, in his essay, *The First Fall of Snow*. 'There is not one of us,

I suppose, who does not experience a curious sensation of romantic interest at seeing each winter the first fall of snow.' He goes on to remark that even 'the most stolid citizen upon noticing white flakes on his sleeve is disposed to raise his eyes and for a moment at least to contemplate with a vague, uneasy interest the clouds above his head.'

Snow! Contained in the word are excitement, joy, beauty, possibility, change, uncertainty, danger and possibly many other things, things I'll look at in this book.

Meanwhile, winter is on its way, yesterday's snowfall has assured us of that. It may be some time coming in force, there will be no doubt some good days of warm sun first, but snow has sent its calling card, and all we can do for now is await its formal arrival.

A Little Science

Twenty centimetres of snow has fallen in the night. It's my first job every morning to clear a path across the balcony outside our bedroom, to the post where the satellite dishes are fixed. Having chosen to live somewhere without a landline and with intermittent mobile phone reception, being able to connect to the Internet is important if we want to keep in contact with anyone. The dishes are surprisingly tolerant of snow accumulating on their reflectors, but even a couple of centimetres on the antennae are enough to disconnect us from the outside world. There's a curious and perverse pull in that idea; if we just allow the snow to collect. . . for a short time the rest of the world ceases to exist.

This morning, as I shovelled the snow over the edge of the balcony, I paused from time to time to watch a small phenomenon that pleases me. As I dumped the snow on the top of the slope beneath the house, little skittering balls ran away down the hillside, making what are known as snow wheels. These only form when the snow is just right; and this morning is one of those times. As a tiny clump tumbles down it picks up more snow as it goes, growing, forming a spiral wheel of snow, that might either stop if it hits a flatter patch of ground, or run on down through the forest to the deep gorge below.

Snow wheels are occasionally made by clumps of snow falling from the heels of walkers' boots; trudging across a sloping terrain of snow, these little Swiss rolls can sometimes form, scurrying away down the hill. On very rare occasions, if the wind attacks a slope of snow at the perfect angle, they can even appear by themselves, where they stand like weird ephemeral monuments, made not by prehistoric hands but just by the action of the natural world. Sometimes, a single starting point triggers multiple streaks of wheels racing down the slope, intertwining, weaving, twisting, even running back uphill a short way as they slew to the side and die. The tracks formed in this way are beautiful pieces of natural art, reminiscent of the traces of exotic interstellar particles caught by the detectors of bubble chambers like that at CERN.

Anyone who's ever been out in the snow and messed around will know that sometimes snowballs and snowmen are easy to form, while at other times the snow just will not behave and falls apart like sawdust, and this is because there are many different types of snow. How many? It's a question akin to asking how many types of leaf there are, because it's not just a question of enumeration, but of classification too. Better not go in that direction; not if you like nice neat answers. Instead, it's worth knowing a little bit more about what snow is, how it forms and the varieties that form can take, and the resultant effect on us.

Why does this matter? I think it matters because we live in a world of over-simplification. Few people have the time, energy or desire to see the world as any more complex than they can cope with. Anyone of my age and nationality will remember the condescension and sneers that greeted Britain's national rail company when they announced that trains were delayed because there was 'the wrong kind of snow' on the tracks. Except that's not what happened. In fact, British Rail's Director of Operations stated they were 'having problems with the type of snow, which is rare in the UK'. In response to which Jim Naughtie, interviewing, replied, 'Oh, I see, it's the wrong kind of snow' and that is the phrase that was adopted in many newspapers and so entered the language as another form of snowclone. 'The wrong kind of X' has been reprinted in various forms in the national press since the original incident, almost always in connection to our train services.

People hear what they want to hear, what they have the energy and desire to hear, and that matters a lot when we come to a bigger debate; the one about the future of our species, and whether that species has irrevocably altered the climate of the Earth. Things are almost always more complex than we would like them to be, and that there are very many different types of snow is obvious, here in the mountains. In a department where the locals are used to heavy snowfall, and know how to drive in such conditions, there are still days when despite the fact that your car is four-wheel drive and fitted with winter tyres, you feel the vehicle start to creep sideways. When you live on a road that has a sheer drop of a couple of hundred metres to one side, and without a safety barrier, such moments are educational. They teach you to stay at home.

One Sunday last winter we woke to find over fifty cars abandoned on just the stretch of road visible from our house, leading to the ski resort further up. The roads are busiest on Saturdays, being changeover day for most ski chalets, and that there had been mayhem on the road overnight was obvious. Amongst the cars were numerous expensive four-by-fours, whose drivers had encountered the 'wrong kind of snow'. As it happened, it was not an especially deep fall of snow that caused problems for British Rail back in 1991. It was a just a very powdery type of snow, one that blocked air intakes on engines, collected in points and sliding doors, causing them all to fail.

So, to ask a very fundamental question, what is snow?

I have to keep reminding myself that snow is not a fifth element. It's easy to feel this way on a day when, in just a few hours, or even minutes, the landscape is transformed before your eyes. The snow has been late in coming this year. After last year's 'warmest December on record', we have just had another one. There was an initial heavy fall in November, and a flurry or two before that, in October, after which summer seemed to return; endless days of clear blue skies and warm sun. On December 11 we ate outside at our favourite restaurant down in town; I had to take my coat off because I was too hot. Christmas came and went, and then, four days ago, the snow finally came.

As much as I would like to see it that way, snow is not a fifth element, but in a single afternoon, a late autumn landscape became a deep winter one. It was made so by snow; not an element in its own right, of course, though I feel it's perhaps permissible to call snow an elemental power.

Snow is just one of the forms that water takes. A snowflake is nothing more or less than a crystal of ice, of frozen water, but one of specific origin, and the nature of snow is entirely dependent on the remarkable nature of water, a substance with many unique properties. A little time thinking about water will aid our thoughts about snow.

To begin with, water is the only natural substance found in all three physical states (solid, liquid, gas) at the kinds of temperatures that occur naturally on Earth. We know water not just as water, but as snow and ice, and clouds and steam, and all are familiar to us. Compared with what we would expect for similar molecules, water has a very high freezing point, as well as a very high boiling point, and big temperature range between these two. If 'Hydrogen Oxide', H_2O, behaved like the other hydrides related to it, we would expect it to boil somewhere at *less than* -62°c, not at *plus* 100°c. We could expect it to freeze not at 0°c but at somewhere *less than* -84°c. But it does not; instead we have that big range of 100°c between these two critical points, and that's just as well, because life on earth would not be possible as we know it. Small changes in temperature would make water either freeze or boil, and pretty much all organisms as we know them would be unable to live.

One final thing, critical to the discussion here, is the behaviour of water as it cools. Recalling school science lessons, we know that things get less dense as they become hotter, and more dense as they cool. Following that to its logical end, ice should be *denser* than water. If that were the case, ice would sink in water, but a moment's thought about ice cubes in martinis or ill-fated ocean liners shows us that this not the case. The reason it doesn't is due to another of water's unusual properties; like most substances it does become denser as it cools, but then, when it slides below 4°c, it becomes less dense again. In fact, water as ice is about 9 per cent greater in volume than it was as water at 4°c. This is why ice floats on water, and why if you put a glass bottle of water in the freezer it will most likely shatter.

Water, as we know, is H_2O – two atoms of Hydrogen combined with one of Oxygen, and the properties described above are a result of the fact that the bonds in the water molecule are very strong – they require a large amount of energy to be broken – and the fact that water molecules can combine with each other in various useful ways. As water cools, the molecules of H_2O are able to align into the open, crystalline hexagonal arrangement of ice.

It is this hexagonal structure that is the seed of the usual six-sided symmetry of the flake, though this hexagonal form is not the only one ice can take. (It can also form a cubic pattern, having a structure similar to diamond, but this only occurs at very low temperatures, below -140°c.)

So far, things are straightforward enough. Water vapour in the atmosphere freezes and forms an ice crystal. (In order to be snow, the water vapour must turn straight from its gaseous state to ice. If it condenses to rain first, what falls from the sky is sleet.) The crystal of ice, being heavier than air, starts to fall to earth. As it does so, it grows out around its six-sided shape, making the familiar arms of the snowflake. The amount of ice that forms at every moment of its descent varies, depending on various factors such as the air temperature, pressure and humidity. Since every snowflake must take an individual path towards the ground, no two will have had precisely the same conditions for growth, and so no two will be alike. We're told from the age we first go out in the snow that no two snowflakes are identical, and this is why. That's not to say that theoretically it would be impossible to find two matching snowflakes, just very unlikely, and in fact it is possible to grow identical snow crystals under carefully controlled conditions in the laboratory.

(Incidentally, still thinking about common beliefs, there is little truth in the old saying that 'it's too cold for snow'. It can be too dry for snow, however, and because the colder the air is, the dryer it tends to be, there may be some correlation between extremely low temperatures and decreased snowfall. But even the coldest air holds some humidity, so in theory it can never be too cold for snow.)

So the structure of snow seems easy enough, but things get more complicated, for the six-armed star shapes, known as dendrites, are not the only kind possible. It's also possible for snow to fall as needles, columns, hollow columns, six-sided prisms and plates. Dendrites themselves vary considerably, and can even come in twelve-armed forms. If a snow crystal falls through droplets of water, these can freeze onto the surface, and are called rime. If the crystal becomes completely encrusted with rime, what falls from the sky is called *graupel*, which as we saw earlier is a thing distinct from hail, which is larger and more solid.

Triangular plates are also possible, as are 'diamond dust' crystals. These latter typically form in extremely cold conditions and produce a beautiful sparkling in the sky as they fall. I witnessed this phenomenon once in Northern Sweden, far above the Arctic Circle, and it is a genuinely magical thing to see. If you're extremely lucky, it can create sundogs. Sundogs are phantom spots of light on either side of the sun, and can be caused by diamond dust in the lower atmosphere refracting the sun's rays as in a prism.

From this brief summary it's easy to see that snow is much more complicated than we might at first think, the interaction of this great number of snow types with each other only leading to ever more complexity. And that's just to describe what happens in the air, as snow falls. Once snow lands, drifts, accumulates, thaws, refreezes, slides and so on, it develops even more intricacy, even greater wildness, but that's another matter again.

For now, it's impossible to avoid a matter about which the scientific community has had much to say, one that lurks in the minds of many people: the changing nature of our climate. There seemed to be more snow in my childhood; that is something I have believed for a long time now. But were the winters worse then? Have we entered a new world of climate and with it shifting weather patterns, which will never return to how we once knew them? Or, if only our memories were longer, would we see extremes of both severe and mild winters? Ask the locals here in the Haute-Savoie and they will tell you about both kinds; this year, we're told, has been the warmest winter in more than sixty, and they said that last year too.

Our memories are not good. No matter how much we like to think we have strong and accurate power of recall, the evidence of psychologists suggests otherwise. We easily fabricate, recast, adapt and destroy our memories, and one of the greatest obstacles we face when trying to keep accurate memories is time. Here, I don't mean that the longer ago something occurred, the harder it is to recall. That can be the case, but everyone can remember in explicit detail something that happened to them when they were five, or nine, or whatever. Some moments survive the impact of time, and we recall them 'as if it happened yesterday'. What I mean is this: time has strange powers, ones we are more or less aware of, but which we still struggle to understand. The way time telescopes away from us is something very strange indeed. As we get older, we all feel that sense that time is moving faster in some way; that years fly by, while, when we were very young, they seemed to take forever. These perceptions have been well-established by psychologists, but I think they apply into the future as well as into the past. We live in a *two-sided* telescoping of time; presumably because we are not only egocentric, but that egocentricity makes us temporal-centric too. What matters to us is what happened just yesterday, now and what might happen tomorrow. Paradoxically, memory, which we usually believe to be about the past, actually evolved I believe to help us deal with the future. So we live in a near-sighted time bubble that seems in focus, while the distant past and far future are 'another country'.

All this makes us very bad at judging questions such as: did it snow more when I was young? Were the winters colder then? This is a sense I have always had; that we were *always* being stopped from getting to school by snowfall in my childhood, not just once or twice. It's so easy to simplify, it's so easy to trust one's instinctual answer and 'know' it must be right, but life is always more complicated than we would like it do be.

In order to find out one small piece of the truth, I contacted the Met Office to find out the real picture, for one young boy growing up in the south-east of Kent, at least.

I was born in 1968, in a small village six miles from the English Channel. I remember my mother often spoke about the infamously hard winter of '63. My parents were living a little way to the north then, on the coast, and she told me how the sea froze that winter, something I had to wait over thirty years to see for the first time, in Estonia.

But did it snow more when I was young, or is it just my desire and recreated memory? To find out, I asked the Met Office for records of days of snowfall for the Manston weather station, just a few miles from where I was born and grew up. I threw them into a spreadsheet, I plotted a graph. I can honestly say I was expecting to be proved wrong, and that it snows just as much now as it did then. I was shocked at how right my original feelings had been. The graph in front of me now plunges steeply down, in just the years from 1961 to today. Yes, there are fluctuations of highs and lows, but the trend line is sharply down, from an average of twenty-nine days of snow at the start of the 1960s, to just nine at the end of the century.

For interest, I decided to look further back. I'm almost fifty, and that's one kind of lifetime, but to look at a slightly longer lifetime I went back to 1934, the year my mother was born. I couldn't get snowfall figures for these years. Instead I looked at the average maximum and minimum temperatures for each month since November 1934, again for the Manston weather station. The average minimum temperature in that period has risen about one degree Celsius, the average maximum about two degrees.

This is all accurate in its own way, though not particularly rigorous, and yet, for one small corner of the world at least, the south-east of Kent, the world is getting warmer. Not long after the phrase 'global warming' was coined, scientists realised it would be better to call it by another name. Today we tend to say 'climate change', simply because its was all too easy for those people who deny the global, numerical evidence to say for example, well, the UK just had two of its hardest winters on record, in 2009 and 2010, so how can the world be getting warmer?

And this is what I mean about simplicity of thought.

The world *is* getting warmer. That one or two degrees increase recorded at Manston, reproduced globally, could be enough to melt sufficient of the ice caps to flood Holland and various parts of Asia. Estimates vary on the metre-per-degree formula; it's maybe enough to know that two degrees is, well, an awful lot, enough to give us serious concern for our future.

But this doesn't mean that it is a smooth process, this doesn't mean that there aren't fluctuations and that within these fluctuations there can be extreme weather events of all kinds: droughts and floods, record lows and highs of temperature. So the term climate change better expresses what we experience in the real world, and will continue to do so. The only argument left is whether we, mankind, are responsible for this change, and if you want to deny that, then place a graph of rise in global temperatures against a timeline of the growth of industry, and then tell me it's not our doing. Don't take my word for it; the IPCC (Intergovernmental Panel on Climate Change) were able to conclude with greater than 90 per cent certainty that man-made greenhouse gas emissions caused most of the observed global average temperature rise since the mid-twentieth-century.

At some point in our future then, there will be no more snow, no more ice. It will exist only in a fake version, in the sterile, indoor playgrounds of the super rich, on ski slopes in hideous shopping malls, reused and artificial snow, like breath that has passed through too many lungs. Real snow – fresh, natural, ephemeral and almost supernatural – that will be gone. Icicles like dragons' teeth, lakes to skate on: these will be gone too. The temperature of the world will have risen to the point where such things will live only in the memory of those old enough to remember, and then snow will take on itself an even deeper symbolism; it will become even more magical, mystical. It will stand then for what we have lost.

For now, the complexity of climate change means that

warmer air provides more 'fuel' for storms, creating stronger weather systems that can lead to greater localised snowfall.

In the time I have been writing this chapter, the snow has not stopped falling. It's just above zero and the snow is heavy, almost slush. It's the kind of snow that hangs heavily on the fir trees outside my window, that can even make a tree that is not strong enough bend over, touch the ground, unable to bear the weight. A while ago I noticed the tell-tale orange light on the satellite modem, and went down to the balcony to clear the snow. Another twenty centimetres has fallen. The most accurate forecast we use says the snow will keep falling for the next six days. That's as far ahead as it goes. That's as far ahead as my 'future memory' wants to go right now. Dig the snow, keep the stoves alight, wait for the sun to return. That's enough.

The Art of Snow

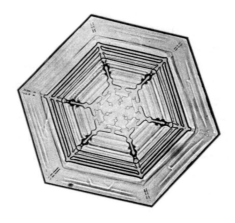

For now, the snowfall has gone, replaced by heavy rain. It washes the trees of their heavy burden, allowing their branches to lift again in an upraised gesture. When I was young, I could never decide when drawing a Christmas tree whether the branches should point down or up. It took me years to realise it depends on whether the tree has snow on it or not.

It's late January, and it's too warm. The temperature has been hovering a little above freezing for a few days now. At times, the heavy downpour has thickened, and become freezing rain, or what we in the UK call sleet. (In some parts of America, sleet refers to ice pellets, and what we call sleet is sometimes termed freezing rain.) In Commonwealth countries at least, sleet is a mixture of rain, snow and partially melted snow, collecting on the ground as grey slush if it falls sufficiently heavily.

Once or twice in the last three days, the sleet has become thick enough to be considered snow, but even an hour of that is soon washed away by warming rains, rains which caused the half metre of snow that had accumulated on the tin roof to slip and slide, making the house rumble as if thunder was overhead. Since we live in an area of moderate avalanche risk, it's a disconcerting sound to hear the first time it happens.

As it happens, our chalet was built in a spot that is supposedly safe from avalanches, though all around are high-risk areas. Friends of ours a couple of kilometres further up the valley live in a house that was rebuilt in the nineteeth-century, after the original was wiped away. When the house was rebuilt, a small hill of rocks was built too, in the path of any future snowslide, hopefully strong enough to do the job, if ever called upon.

For most of us, life in the fullest sense of the word is unthinkable without some form of art. I have met people who deny there is any great need for art; that the important things in life are food, shelter, education and so on. And yes, of course these things are vital, but without art in some form, be it music, film, literature, etc., we are not living. We are merely surviving. Art is what helps us interpret (not to mention enjoy) the world around us. I hesitate to say 'understand', because I wonder how much we ever fully understand. But in trying to make sense of life, we continue to pour out an endless stream of stories, songs, drama and so on, and have done so, it seems, for as far back as we are able to discern.

The natural world must have been one of our earliest sources of artistic inspiration. This is something I have become aware of as I developed as a writer. I'm only about twenty years into a writing life, but I can see I have become more self-aware as I have continued to work. When I started, I just wrote, without thinking beyond the surface of what I was doing too deeply. But as the years and the books went by, I found it impossible not to reflect upon what I was doing, and why. This can be a good thing. I also think it can be a bad thing. Several times, I have found myself in a period of writer's block, stymied by over-complicating and hampering thoughts. At times, I felt I would never know how to write anything again, unable to see a way past mere repetition of what had gone before.

Eventually however, the light shone again and work recommenced. Maybe I just got too desperate not to work, but each time it's happened I've come away with a slightly better understanding of what I'm doing, and how.

Years ago, with one of my early books, I wrote a scene of snow. In it, the protagonist, who had just lost a beloved protector, was standing at a window, watching a long-awaited snowfall. I described how he stared at, as well as through, the snow, how he might follow the final moments of one individual flake as it fell to its end, and how, as the snow accumulated on the ground, it began to soften his grief, to numb it. It's not the best scene I've ever written, nor the best book, but I mention it because it was as I was actually writing it that one of these moments of self-awareness occurred. In the past, I might have written this metaphorical scene from instinct. Now, I would be very much aware of what I was doing, and how I was using the snow. As I said above, this is not necessarily a good thing. I'm reminded of Charles Dickens' supposed last words 'Be natural, my children.' I have no idea whether he actually said this. That doesn't matter. The full quote is this: 'Be natural, my children. For the writer that is natural has fulfilled all the rules of art.'

How can we be natural when we are overeducated in our art form? This is the crisis underlying the failure of much modern art and writing. We are too wise, too cynical, too knowing; we feel that everything has been done. In an attempt to push on regardless, we innovate pointlessly, without honesty. We are no longer natural.

How much easier then, thinks the modern artist, to have been an original! How much easier to have lived during so-called golden ages of music, or art, or literature. To be influenced not by forms, or critics, or peers, but simply by the world around you, things, for example, like nature, like snow.

Peter Bruegel the Elder lived during the time known as the Little Ice Age. Although exact dates cannot be agreed on for this period of global low temperatures (due to regional variations, amongst other discussions), it lasted for a few hundred years at least, only ending in the nineteenth-century. Though there was worse to come after Bruegel's time, he would have been very familiar with the extremely cold winters of the sixteenth-century. These were days when rivers and lakes froze for weeks on end, and hunger and famine were prevalent. With *The Hunters in the Snow*, painted in 1565, Bruegel produced a painting that has been called the most famous depiction of snow in Western art history. It has also frequently been cited as prima facie evidence of the Little Ice Age itself, though the landscape that Bruegel depicts in the painting is not exactly the Netherlands.

The picture shows a winter scene. A group of three hunters return from an expedition with their dogs, on a hillside above a valley. Heavy snow lies on the ground all around; but none is falling, the sky is a brooding icy green, the trees rimed with frost but not laden with snow. It appears to have been a similar story to that which I describe above – a good deal of snow has fallen, but some thaw or rainfall has washed the branches of the trees free of it.

Bruegel captured the mood of a winter that has long set in, that has been hard and cruel; you feel the cold looking at the painting. And Bruegel will certainly have known harsh winters, yet he chose not to represent his home landscape, or if he did, not with any great need for indelible accuracy. For while the valley towards which the hunters trudge shows skaters on what appear to be man-made lakes, surrounded by buildings of Flemish architecture, there are no mountains like the ones depicted here to be found in the Low Countries. The hill in the foreground on which the hunters are forever frozen is high enough; the towering mountains in the distance would be more at home in the Alps. In fact, we can assume this is where they came from; after a spell in Italy, Bruegel returned to his home in Antwerp around 1555, by way of the Alps. He made many drawings of the mountains, but even here was not content to record nature exactly as he found it. Rather, he made fantastical composite drawings of mountain ranges, some of which he used in later paintings.

What Bruegel captures above all else, I think, is the way it feels to be outside in the cold and the snow, with all its harsh raw threat, albeit a threat that is, for now, contained. A threat that lurks under the surface of the snow, just hidden by the elemental power itself.

In the air and in the trees above Bruegel's hunters are black birds. This is not an ornithological work; it's hard to tell what they are exactly. For me, they can only be crows of some kind. These birds, carrion birds, are constant harbingers in many works of snow art.

In 1974, the German film critic Lotte Eisner lay in Paris, dying. Her disciple, the still-young director Werner Herzog, decided he had to visit her. Convinced that Eisner could not, would not die before he saw her, he decided to walk from Munich to Paris to see her. It was late November when he set out, with winter approaching fast. He kept a diary as he walked, and some years later this diary was published in an almost unedited form as *Vom Gehen im Eis*, or *Of Walking in Ice*. The title is a metaphor for life, without doubt; the book itself surreal and dreamlike for much of the time. This is only natural for something not originally intended to be read by anyone else, and yet is what gives the book much of its power. Herzog walked 'as the crow flies' whenever possible, buying new maps whenever he needed one, sleeping rough at times, breaking into barns and houses when necessary, occasionally accepting offers of accommodation and food. Throughout the book he notes the crows that seem to accompany him.

He also describes the weather frequently, almost obsessively perhaps, in tones such as these: 'a black morning, darkened, so dark and cold, a morning that lies on the fields only after some great cataclysm, after some great pestilence.' This is understandable. Walking in the landscape, it is impossible for it not to assume different proportions from those as viewed from behind a window, or behind the wheel of a car. I recall the writing of Joseph Armistead, whose book *Ten Years Near The Arctic Circle* described his work as a missionary in Norway in the early twentieth-century. 'O ye, who live in luxurious homes and lie on downy pillows, who ride from place to place in comfortable electric cars, who send for the doctor and have him sitting at the snug fireside in little over half an hour, think of those who live amid these dreary wastes of snow and ice and barren rock.' Quite.

Herzog's account is this: of walking through dreary wastes, with crows for company, towards someone else's death. It mirrors a similar account, though this time, the death our wanderer is walking towards is his own. I'm referring to Wilhelm Müller's poem *Die Winterreise*, and Franz Schubert's transformation of it into the epic song cycle known even more simply as *Winterreise,* meaning Winter's Journey.

I find *Winterreise* to be one of the most devastating works of art, of any kind. It is certainly the work of art that most powerfully conveys isolation and loss in a vast metaphorical sweep of snow. Müller's poem cycle tells the barely explained story of a young man who has been rejected by his lover (or are we to assume fiancée?) and now wanders the winter's landscape, in search of... what? Perhaps the poet-narrator knows what he is searching for from the start, perhaps that only becomes clear to him as he wanders towards the final, fatal conclusion. Schubert, still a young man, but almost certainly already aware that he was slowly dying, took Müller's poems and turned them into a monster of some kind.

Today, a *Schubertiade* is a gathering of musicians performing works by, or inspired by, the Austrian composer. These affairs began as little salons at his patrons' houses, for his friends and benefactors, often to trial new pieces. *Winterreise* falls into two halves of twelve songs each. The story goes that Schubert called his friends together, one evening in 1827, to hear him perform the first half. When he finished, there was stony silence, disbelieving faces. What was this utterly bleak and strange music that their hero had composed? They were dumbstruck by the weirdness and terror of the piece, and yet the first half, the half they'd witnessed, is nothing to the second, something so completely desolate it makes the first twelve songs appear like the dream of a summer's picnic.

What was it that Schubert had captured? Why is it so powerful, even today? One of the notable friends who attended the original Schubertiades was the poet Johann Mayrhofer. He described Schubert's mood at the time as one of deep melancholy: 'life had lost its rosy colour and winter was upon him.' Schubert was painfully (and I mean it literally) aware of the state of his health, and that time was not on his side. Although the official cause of his death would be recorded as typhoid fever, the likely truth is that he was suffering from the tertiary stages of syphilis, the treatments for which at the time could do more harm than good. Mercury, in the form of pills and ointments, was one very common method for attempting to remedy the disease; it is known that many of Schubert's symptoms match those of mercury poisoning. He would not have needed a doctor to tell him that the prognosis was not good. Death is indeed all-too-present throughout *Winterreise*, and when we learn that Schubert was still editing the proofs of the second twelve songs just before his death, we do not need to wonder why. He died a few days before the final publication of the song cycle. The short life of this incredibly productive composer was over, and he was not yet 32.

And yet there is hope, great beauty and even a little joy in the songs too. From the first half, *Der Lindenbaum*, The Linden Tree, has grown to be the most famous of all the songs, perhaps second in fame in Schubert's oeuvre only to *Die Forelle*. It is often performed as a solo work. Together with *Frühlingstraum* (Spring Dream), *Lindenbaum* is maybe the sweetest of the twenty-four, and is the first one that begins in a major key, but even here the beauty is stabbed, twice and hard, by the minor. For example, the lines which describe how the tree seems to call out, offering memories of better days, and maybe even salvation – 'come here to me, old friend, here you'll find your rest' – are immediately dispelled by 'the cold winds blew straight into my face, my hat flew from my head. I did not turn back.'

There may seem to be joy here, but there is not; there is only illusion.

There is illusion too in another happy-sounding song, *Täuschung*, towards the end of the cycle. Täuschung means 'deception' – the music may sound cheerful, but it only provides greater juxtaposition to the bitterness of the lyrics, and when this short and lovely song is over, we embark upon the final, five, shattering songs of the cycle, in which our hero stumbles across a graveyard yearning for death, and yet, even here, comparing the graveyard to a crowded inn with no rooms to spare, he finds rejection.

This is the nature of *Winterreise*. It lures you in, with elegance and beauty, like a fall of snow, but waiting for you at the end are the inevitable consequences: the frostbitten horror of the snowstorm, as the poet-wanderer meets his death in the white wasteland.

There are also crows, and ravens, following our wanderer just as they followed Herzog on his epic walk to Paris, just as they lurk above Bruegel's hunters.

> *Eine krähe war mit mir,*
> *Aus der Stadt gezogen.*   Pulled from the town,
> *Ist bis heute für und für*   a crow was with me
> *Um mein Haupt geflogen.*   flying over my head,
>   day by day.

The winter wanderer sings to his crow companion, perhaps grateful for the company, but in truth fearing that all the bird really wants is a meal of fresh carrion.

And then, at the close of the cycle, in the penultimate song, come the phantom suns. *Nebensonnen* is the German term for the phenomenon mentioned in the previous chapter: Sundogs, more scientifically known as parhelia, formed by the prismatic presence of diamond dust ice crystals in the atmosphere. In *Die Nebensonnen*, Müller and Schubert's wanderer comes face to face with this otherworldly spectacle, in itself a strange portent: two mock suns hovering either side of the real one. The poet sees the phantom suns as the eyes of his lost lover ('now the best two have gone down') and wishes only that the third one, the real sun, would go too.

We approach the end. With the poet-narrator, we stumble across a lone figure in the white wilderness: the organ grinder, the hurdy-gurdy man, cranking out with frozen fingers a simple tune on his music machine. *Der Leiermann* is not for the faint-hearted. It opens with a two-note discord, mimicking the drone of the hurdy-gurdy, it closes with the singer's plaintive cry to the 'organ-grinder' himself: 'strange old man, should I go with you? Will you play your hurdy-gurdy to my songs?'

Of course, it sounds better in the original language.

> *Wunderliche Alter*
> *Soll ich mit dir gehn?*
> *Willst du meinen Liedern*
> *Deine Leier drehn?*

I don't mean to sound pretentious – it's just one of those works of art that makes you wish you were fluent in the sung or written tongue. If you don't know it, hunt out a recording of Dietrich Fischer-Diskau singing *Der Leiermann*, and prepare to shudder.

That Schubert created such a bleak, such a modern, such an *empty* piece of music in 1827 is to be marvelled at. That he had the courage to write it down in the months before his death is a testament to the grip and lure of the best art. It is one of those works that sounds as if it always existed, as the cliché has it, that it was discovered, not composed. In these short lines, Müller's words and Schubert's music combine to destroy all hope utterly, forever, and leave you with a burning bitter tear in the corner of your eye. Beyond that, there is this: running through the heart of *Winterreise,* made explicit in the final songs, is the true terror of the snow. It is isolation.

Almost exactly a century later, Thomas Mann made great use of *Winterreise* in his greatest work, *The Magic Mountain.*

The Magic Mountain is at once both a prime example of, and a subtle parody of, the Bildungsroman; a novel that depicts the 'education' or 'formation' of its protagonist. As the novel opens, young Hans Castorp travels from his native Hamburg to Davos, Switzerland, to visit his cousin Joachim, who is a patient in a tuberculosis sanatorium. He intends to make a visit of three weeks. He stays for seven years.

Another work that makes me wish I spoke more German, *The Magic Mountain* is a vast work, running to over 700 pages in most editions. In that space, it is a book that can be many things; it is a beautiful, sad, creepy, funny, wise, challenging novel that asks many questions, about time and the perception of time, about illness and death and love, and many other things besides, but as Mann describes life in the sanatorium, it is a very largely allegorical novel. The sanatorium is the world; Hans Castorp (always called by both his family and given name) represents the youth of Germany, as the First World War approaches from the distant horizon, silently.

Two characters in the novel in particular try to 'win' Hans Castorp to their own camps. Naphta, an ex-Jesuit, now Hegelian-Marxist, preaches destruction and war, while the Italian Humanist, Settembrini, who we are often told reminds Hans Castorp in some way of an 'organ-grinder', tries to steer our hero towards civilisation, tolerance, and democracy.

As well as this organ-grinder reference, Schubert's *Lindenbaum* makes many appearances in the novel; for example, soon after his arrival at the sanatorium, Hans Castorp marches up into the mountains, singing the song aloud with no one to hear him. Later in the book, when the management purchase a phonograph for the enjoyment of their 'guests', Hans Castorp takes it upon himself to appropriate control of this modern device and plays various recordings to himself, again and again, one of them being *Der Lindenbaum*.

It is on a second excursion, much later in the novel, that Hans Castorp almost dies. This is the chapter called *Snow*. Writing in his book-length study of *The Magic Mountain*, Hermann Wiegand called this episode 'without question the spiritual climax of the whole novel'. Since the story is set high in the Swiss Alps, there is much mention of snow in the book, but in this important chapter Mann unleashes all his finest word-weapons.

A prolonged period of heavy snowfall has beset the sanatorium. Unable at times even to take 'the lying cure' on his balcony, Hans Castorp cannot fight the urge to escape the walls of the place, and his fellows. He experiments with the fine art of skiing, and despite being no sportsman one day embarks on a long adventure in the snow-clad mountains. He rapidly becomes wildly lost, realising he has passed twice by the same hay hut. Being lost irritates him enormously, and he vows not to let himself be defeated by the snow: 'this idiotically symmetrical crystal geometry.'

Lying down in the lee of the hut for a while, Hans Castorp teeters on the edge of reason, of consciousness, of life itself. In half dream/half hallucination, he witnesses sublime visions of some bucolic idyll, on a warm, sunlit beach, visions which suddenly turn almost apocalyptic. In a temple, some way beyond the beach, he witnesses an appalling scene: two old women dismembering and eating a child.

Very often then, in literature, snow symbolises and brings isolation, and death. Ironically, however, the broken-hearted traveller of *Winterreise* is not alone in his wanderings. He may not be aware of it, but he has been accompanied by many other solitary figures in the snow over the years, often heading towards their doom. Here are just a few.

At the end of *Women in Love*, D.H. Lawrence suddenly pitches the foursome of Ursula and Birkin, Gerald and Gudrun, into a holiday in the Alps: 'the eternal closing in, where the walls of rock and snow rose impenetrable and the mountain peaks above were in heaven immediate... the centre, the navel of the world.' It's here, following a final terrible fight with Gudrun, that Gerald staggers out into the snow, lies down, and 'goes to sleep'.

It's with snow, in polar wastes, that Mary Shelley chooses to show us the dreadful opening and closing device of *Frankenstein*, as Victor pursues his monstrous creation across the most desolate face of the Earth.

It's snow that Steven King and more successfully Stanley Kubrick use in their respective versions of *The Shining,* to demonstrate not only Jack Torrance's isolation and solitary death, but the isolation which we all face, unable to ever truly be part of anyone else. As with Mann's sanatorium, the Overlook Hotel represents the world; when Jack tries to kill his own wife and child therefore, it is nothing less than an act of attempted genocide. He tries to kill a whole people.

Examples of snowy death in art are as endless as the polar wastes where we're heading next; here's one final one that sums everything up before we go there. *Il Grande Silenzio* (The Great Silence) is an underrated gem of a movie. Made in 1968 by Sergio Corbucci, the film was part of the wave of Spaghetti Westerns characterised most notably in the 1960s by the films of Sergio Leone: *A Fistful of Dollars* and its kin. Ennio Morricone, who wrote the music for those pictures, worked on *The Great Silence* too, producing some of his most haunting compositions. Yes, this is a genre movie, but there is more to it than that. Co-written by Corbucci and his brother Bruno, the piece was created as an allegory of the recent deaths of Malcolm X and Che Guevara; this is a Western that overthrows the ambitions of other Westerns. The story is set in deep, deep snow in Utah, in the winter of 1899. (This winter was exceptional by any circumstances; known as the Great Blizzard of 1899, winter conditions reached as far as Florida and Louisiana, the port of New Orleans itself even becoming iced over.)

While most Spaghetti pieces were part of a growing trend to overturn the simplistic moral notions of the Western before this point, Corbucci's movie moves things up a level. In the opening sequence, the eponymous hero of the film, a mute well played by Jean-Louis Trintignant, is greeted by a murder of crows who take to the wing, startled by the presence of bounty hunters out for our hero's blood. When they draw their guns, he takes them down, all apart from one who surrenders; 'The Great Silence' spares his life, but shoots off his thumbs so he cannot use a gun again. And that's our 'good' guy. The 'bad' guy, played in usual menacing style by Klaus Kinski, is the most politely spoken and apologetic killer to grace screens until Kevin Spacey turns up in *Seven*. 'Try to understand, Madam,' Kinski says to the aged mother of a fugitive he and his accomplice have just shot in cold blood, 'it's our bread and butter.'

The film moves on through the snow, violently and remorselessly, even if some of that snow is FX foam and not the real stuff. But much of the movie was shot on location in the Italian Dolomites, and this landscape combines with Morricone's memorably melancholic music to offer the true message of the film, as not only does 'good' fail and 'evil' triumph, it does so in the most appalling way possible. As British director Alex Cox said, it has 'the worst ending of any film, ever.' Worst, not as in badly-made, worst as in bleak. The ending was so 'bad' in fact, that an alternative 'happy' ending had to be made for release in certain markets. I've seen this ending, and while it makes you feel better about life, you know it's a lie. It's not true to the story that's being told; and that story is one that says good does not always win, and we must be on our guard for that. The original ending of the film is the one you knew was coming all along, for what else does all that snow suggest other than the futility of facing a power greater than yourself?

As I write, the rain has gone. The temperatures have

dropped and snow is falling heavily, slowly, ponderously. Powerfully. Today is one of those days when it's hard to remember that most of the time, snow does not fall. Today is one of those days when it appears that is has been snowing since time began, that it will keep snowing until time ends, that snowfall is the only thing in the world. It's hard not to be caught by it, hypnotized, and spend the day staring at the world disappearing under the blanket of white. The landscape around our house is very much like the Dolomites where *Il Grande Silenzio* was shot, and watching the snow through my window now, it's another lesson from nature. The snow will stop when it wants to stop, and there is nothing we can do about that. Best not to try.

Exposure

So much for how we view snow from the (relative) safety of our imagination. It's quite a different thing to actually be out in the stuff. As I write this, we are cut off from the outside world. The first thing we found this morning was that a snowplough had managed to do something that has never happened to us in five winters. Our house lies just below a road. Normally, the *deneigeurs* (the workers who toil endlessly clearing the roads in the winter) manage to neatly pile the snow up at the side of the road, without sending too much of it down onto our property. This morning, we woke to find that the twenty-metre path from our front door up to the road had been buried under a metre of ploughed snow.

The reason this had happened was, once again, due to the type of snow – being extremely light and powdery, it had slid down the slope much more easily than it otherwise might. The silver lining was that this also made it easier to clear than if it had been the wet, heavy or icy stuff.

All this was just the overture to discovering that a small avalanche had hit the road down to town. There was not a great deal of snow, but a few medium-sized pines had dislocated the protective barrier of concrete blocks, spilling them across the road, which has been closed until further notice. It's the same spot that was hit last spring by a landslide.

Last year, the transition from winter to spring was characterised by heavy rainfall, which led to some severe flooding in the valley. Up here, in the mountains, the problems were of a different kind. Many of the roads wind their way through deep gorges containing plunging torrents; beside the narrow strips of tarmac vast sections of the rock walls spend their lives chained behind protective nets of steel, which can turn to phantasmagorical curtains of ice in the winter. Last year, extreme water flow was enough to cause serious erosion; a landslide blocked our road down to the flatlands for six weeks, striking at the same weak spot that has just cut us off again, overnight. The Mayor, who has been in office for twenty-six years, told us it was the worst incident he could remember in his time. It remains to be seen how long it takes to unblock the road this time, more serious in the winter because the only other two possible routes out are buried under ten metres or more of snow.

The destructive deluge of water that occurred last spring is made more powerful from the combination of the rain with the melt water from the season's snowfall; the snow essentially creating a reservoir of water held in stasis on the mountains until freed by rising temperatures and the rapid melting effects of rain. This process is a matter of concern in many parts of the world – it is estimated that as much as 75 per cent of the American West is supplied with water that derives from the melting snowpack. The increasingly severe droughts of California, for example, where decreased snowfall in the Sierra Nevada over recent years has exacerbated water shortages. It's also critical that it is snow that falls, not just rain, for beyond a certain point, large deluges cannot be handled. Science writer Michelle Nijhuis notes 'As in most of the rest of the American West, fortunes depend less on how much precipitation falls from the sky than how much of it falls as snow and how long that snow stays in the mountains.'

In California, the movement of water into otherwise infertile areas is a relatively recent, man-made phenomenon. Back here, in the Alps, the effects of the presence of snow is written into the land itself. The gorges, the powerful torrents in their deep rumbling bellies, the glacial origins of the major valleys themselves: whatever the season, this is a landscape defined by snow.

Snow, of course, is not to be taken lightly.

A few years ago now, I made a couple of trips up to the Arctic Circle to do research for a novel called *Revolver*. On the first trip, I flew to Kiruna, a town in the far north of Sweden, and made it my base for a week or so. It's a strange place: a mining town that grew out of next to nothing a little over a hundred years ago. Since then, the deposits of iron ore have been almost mined-out; when I was there plans were being mooted to move the whole of Kiruna, building by building and house by house, a few kilometres east, to allow access to further deposits under the town. The work has now begun, as old buildings are demolished and new neighbourhoods constructed. They've given themselves a decent timescale to complete the work of moving the town of 20,000 people: about eighty-five years.

One February evening I took a flight up from Stockholm and arrived at Kiruna airport. I approached the solitary young man at the car hire desk, which looked very much to be one of those folding tables on which you lay wallpaper out for pasting. I gave him my name; he looked confused, then dismayed.

'There was another guy called Marcus, just got off your plane. I thought he was you so I gave him your car.'

Then, his face brightened.

'Don't worry,' he said. 'Don't worry. I can lend you my van.'

We shook hands, I grabbed my bag, and he went round the back of the large shed that was posing as an airport terminal to fetch his van. What appeared was a four-wheel drive, silver van, with lowered suspension and tinted windows. All in all it looked very much like the kind of van a small rock band would tour Germany in. But there was no choice, and it was a sturdy vehicle, with spiked metal tyres for maximum grip on the ice roads. I made long trips in that van, driving to Finland and Norway in it, and in all that time I could only pick up one radio station, which played nothing but Norwegian Death

Metal. I think it was the van's choice.

One bright day, heading towards Norway, I parked the van beside the road to photograph a mountain range. Trying to get a better angle, I left the verge and explored a short way into some scrubby trees standing nearby. My boots sank a little way into the snow, and I hadn't noticed yet how the trees seemed to be buried up to their lowest branches. I took another step and found out why. I'd been walking on a crust, under which was a metre or more of looser stuff, again, the precise behaviour of snow depending on the precise nature of the fall, thaw and refreeze that had happened. I was up to my chest, and stuck. I tried to get myself out; I could not move my legs.

At first, I didn't panic, the road was a few metres away, and is the only route from Kiruna to Narvik. Someone would come by soon. But they didn't. Eventually, and taking a very long time, I managed to dig my way out, handful by small handful. I sat in the van, with the engine running for warmth, and wondered if I'd been lucky, or unlucky. In the end, I decided I had been lucky. Not only was I still alive, but I had found a useful plot device, one I used in the denouement of *Revolver*, in which the young protagonists are able to escape across a crust of snow while their lumbering pursuer is trapped. I don't believe you have to visit a place in order to convincingly write a book there, but it can certainly help, simply because you very often find things you didn't know you were looking for in the first place.

Showing that we don't always learn from our mistakes, an even stupider thing happened to me (or maybe I'm just the stupid common element here) in the winter of 2010/11. It happened not in the Arctic Circle, nor in the Alps, but close to my cottage, near Cambridge. This winter, and the one before it, were unusually severe for the UK. I have a photo I took of the temperature gauge in my car, which showed minus 10°c, in the middle of one December day. And indeed, that month of 2010 holds the place for the coldest December on average since Met Office records began (And please let's remember to speak of 'change', not 'warming'). My brother and I took our teenage children ice-skating on the Ely Fens. It was a once or twice in a lifetime thing: a mysterious and magical day of mist and deep cold, silent apart from the scrape of the skates on the waterlogged, then frozen, fields.

The record for the overall coldest winter in England, by the way, remains that of 1684, but then, that was during the Little Ice Age. How do we know this? Well, there is such a thing as the Central England Temperature record, preserving monthly temperature records to at least the nearest half-degree, unbelievably, since 1659.

Back to 2010. When I lived near Cambridge I often walked from my village to a friend's house in the next village, about a mile away. He was having a dinner party, as he did often, and in those days I could easily walk the mile or so, and would do so whatever the weather. I would enjoy the freedom to be able to drink and walk home, happy not to be reliant on the car, as you frequently are when you live in the countryside. It had been snowing; it had been snowing *a lot*, and being near the edge of the flatness of Cambridgeshire, the snow had been drifting. Around seven o'clock, I made my way out of the village and along the path that leads across the fields to my friend's house. It was heavy going, but I was well dressed in a coat I bought the first time I went to Iceland, decent gloves and warm snow boots. I didn't realise at first how hard the going was, or how tired I was getting. I think it was this that stopped me from realising how deep the snow was becoming in places. Eventually, I found I'd walked myself into a deep drift, and into exhaustion. I gave up, and let myself fall backwards in the snow, remembering that time in Sweden, half-laughing, half-cross with myself. I wondered if it was possible to die from getting stuck in a snowdrift, half a mile from one of two villages in Cambridgeshire, in the twenty-first-century. I don't *think* it is. But, as I arrived at my friend's, a little late but otherwise fine (and admitting to no one what had happened), I decided it's not something I want to prove, one way or the other.

It's said that when you freeze to death, the last thing you feel is incredible warmth. I used to wonder whether those sorts of assertions are true; how people know such things. It is apparently true however, and results from a combination of two factors. First, the effects of hypothermia are many. As core body temperature drops, shivering becomes uncontrollable; the sufferer has trouble with movement and coordination. But also, as hypothermia sets in, confusion is another common symptom; people act strangely, as if dazed or drugged. The body, meanwhile, implements vaso-constriction, whereby blood flow to capillaries is shut down. This is what is responsible for frostbite of extremities, but the body does it for a good reason – it's attempting to maintain the core body temperature of the vital organs. Better to lose a finger than freeze entirely, that's the cruel trade-off.

As hypothermic reactions continue however, the muscles

that enable vaso-constriction become exhausted, and relax, allowing a sudden rush of blood to the skin once again. This causes a sudden feeling of heat, which combines with the disorientation the victim is already suffering from, causing them to undress. If alone, death is now inevitable, though these final stages can also be accompanied by another strange behaviour – that of burrowing. This process is not so well understood, but may be the result of a final primitive autonomous action of the brain, effectively an attempt to dig and thereby create shelter, like an animal going into hibernation.

There are some people, however, who have a survival mechanism that can help keep them alive in dangerously cold conditions, and prevent frostbite too. Inuit and other people native to cold climates exhibit a capacity known as 'the hunter's response'. Such people have to work with their hands exposed to the cold from time to time, and in order to do so their bodies enable a flow of blood to the capillaries every few minutes, to allow a good working temperature to be restored. The interesting thing is that this is not a purely genetic ability – US army instructors claim it's possible to train your body to respond in this way by repeated short exposures of large areas of skin to the cold, while keeping your hands and feet warm. Apparently, over enough time, this teaches your body to maintain capillary flow even in cold conditions. It's another thing I don't intend to find out.

The Inuit have other genetic adaptations to the cold; they are able to live healthily on their traditional diet, which is rich in fatty acids from fish-eating seals and whales. Research has shown that this adaptation means they do not show high cholesterol levels which could have been expected given their nutritional sources, allowing them to thrive in a cold environment, with a high-meat, high-fat food supply.

Along with their genetic advantage over the average Caucasian visitor to cold climes, the Inuit have developed the best methods for living in and coping with the cold, and such methods can be the difference between life and death.

In the epic race at the start of the twentieth-century to reach the 'bottom of the world', no single story struck me more strongly as a child than that of Scott and his rival Amundsen. As a child I read *The Worst Journey in the World*, the classic account of Scott's doomed *Terra Nova* expedition, written by Apsley Cherry-Garrard, a member of the shore party and leader of an abortive attempt to meet Scott on his return from the pole. I read more about Scott and his rivalry with Amundsen, and I remember feeling two things, very powerfully.

The first was a sense of admiration for Amundsen; at his good sense in using the right equipment and clothing, using techniques suitable for moving in extreme cold across ice and snow. He was a proficient cross-country skier, and he took other skiers with him, one of whom was a ski-maker from Telemark. He developed clothing based on that worn by the Netsilik Inuit of Greenland, made of reindeer and wolf skin as well as Burberry gabardine. He used sledges pulled by dogs, and actuated a plan to kill dogs as food on the push to the pole, while also using supplies of seal meat, as it was known to alleviate the likelihood of scurvy.

Even as a child I felt uneasy about the 'noble failure' tag applied to Scott and his men. I have never comfortably bought into the British folklore of 'we lost, but we did everything in the proper fashion', whether we're talking about polar exploration or football matches. Scott was reluctant to make much use of dogs, though he would later acknowledge their usefulness in the right hands. He also used ponies and motor vehicles, but all of these things were only to reach the polar plateau, from where the rest of the journey would be made on foot, man-hauling the supplies in back-breaking fashion. The ponies selected were poor specimens because Lawrence Oates, who was to have bought them, arrived too late to do so. Instead, they were chosen by a man who knew nothing about horses. The apparent calamities of the British expedition are many; their ultimate death on the return from the pole, having been beaten by Amundsen by some five weeks, is the stuff of legend. The noble death – 'I'm just going outside, and may be some time' – proving just what plucky heroes we British are.

Argument continues to this day about both Amundsen and Scott's methods, about their dubious man-management, about their choices in various aspects of their missions. Amundsen is sometimes portrayed as a cold and calculating man, one who tricked Scott, the rest of the world, and for some time *his own men* into believing they were heading for the North, not the South Pole. The British disdained his technique of slaughtering his dogs, though they themselves would eat their ponies as they died. Similarly, biographies and research emerge regularly either to show what a bungler Scott was, or to refute all such claims, citing other reasons for his failure, for example the extremely bad weather in the final stages of the doomed mission.

Despite this, it is still a strong argument that to survive in a hostile place, it's best to use the methods developed by people who have lived there for centuries, rather than try to force your own will on the environment. It reminds me of a (largely) apocryphal tale from the days of the space race, and how NASA spent millions of dollars developing a pen that could write in zero-g, upside down, in any temperature, while the Russians used a pencil. This story is pretty much an urban myth, but like the Inuit snowclone, it persists because it says something very true.

The second thing I took away as a child from reading about such extreme explorations was this: horror. The thought of wanting to, or at least being prepared to, submit yourself to such appalling and dangerous conditions terrified me. The timescale alone in such adventures is unsettling. In all, the *Terra Nova* expedition lasted over three years. In the hunt to be first to the North Pole, Fritdjof Nansen deliberately sailed his ship, *Fram*, into the Arctic ice shelf, trusting that it would eventually drift westwards, and perhaps over the pole itself. He did so knowing it could take perhaps four or more years to do so. Such feats speak of a different kind of mentality, and after all, to achieve such things is not a common occurrence. I suppose, therefore, that these things terrify most of us, which is why we marvel at and respect those individuals who carry out such expeditions, even if we quietly think they're slightly unhinged at the same time.

Polar voyages and other expeditions in cold climes are legion with tales of hunger, deprivation, loss of toes, fingers and noses to frostbite, but then there are the even darker moments: stories of mental collapse, of madness, and of that greatest of taboos: the eating of human flesh.

In 1845, Sir John Franklin set out from Greenhithe, Kent, in two ships, in an attempt to prove the existence or otherwise of the North-West Passage from the Atlantic Ocean to the Pacific. The ships, the *HMS Erebus* and *HMS Terror* (and who would have wanted to sail on that one?) left England in May. By July, they were spotted in Baffin Bay, heading north, and all appeared well, but they were never seen again, by British eyes at least.

After 1845 and 1846 had come and gone, without word from the expedition, the worst had to be feared. Rescue parties were sent out early in 1848, but they found nothing. In 1850, many more ships entered the hunt for the missing parties, but still without success, until finally, four years after that, a Scottish 'overland man' working for the Hudson Bay Company was the first with news of the Franklin expedition's fate. Over the previous years John Rae had made several expeditions along the Arctic Coast, until, in April 1854, he met an Inuk (the singular of Inuit) who told him about a party of white men that had died of starvation. Shortly after, other Inuit confirmed this story, and some had items such as silver forks and spoons, which verified the tale; as final proof, Rae even bought from them a small silver plate with Franklin's name engraved on the back. The Inuit told stories of cannibalism, and Rae duly reported all that he had found back to the Admiralty.

His reports were not well-liked. Franklin's widow, Lady Jane, in particular objected to the suggestions of cannibalism, and her cause was taken up by no less than Charles Dickens, who denounced Rae for believing the stories of 'a race of savages'. If anyone had been eaten, it was argued, it was the natives who had done it themselves, though why the Inuit would have killed and eaten Franklin and his men when they were perfectly capable of fishing and hunting seals seems not to have occurred to Lady Jane and her propagandist, Dickens.

Later expeditions, such as that by Leopold McClintock in 1859, appeared to confirm the Inuit's story. McClintock found an abandoned boat attached to a makeshift sledge. There were two skeletons, along with all sorts of items, such as guns and watches, as well as many items that seemed incongruous: silk handkerchiefs, towels, books and so on. At another site, McClintock found large amounts of discarded equipment, winter clothing, navigational gear, medicine, but noticeable largely by its absence, was food.

The matter of cannibalism began to be resolved in 1997 when remains of the crew found on King William Island were analysed. From this and other studies since, it now seems certain that the expedition had indeed resorted to eating human meat in order to survive. It has also been established that the crew were suffering from lead poisoning, which would have exacerbated their physical and mental health; side effects of this toxicity can include depression, anxiety and aggressive behaviour.

One can only shudder as one imagines the last days of these men, far from home, with little hope of survival. And yet, according to the Inuit, it was some time after the expedition had foundered that some of the men were encountered. While their stories were largely dismissed at the time, since the discovery of *HMS Erebus* in 2014, it now seems that everything the Inuit had to say was true, including the stories of man-eating. The ship was found more or less where the Inuit had said it had been seen, and in water shallow enough for the masts to be visible for some time, as they maintained.

What's also notable is that, like Amundsen after him, John Rae also made use of native techniques for surviving in the Arctic, wearing native clothes and hunting native food sources, and in doing so survived his travels, whereas the Franklin parties perished.

It should not be thought that these voyages into snow are only a result of Man's need for new trade routes, or the desire to conquer new continents. Nor that they are a feature of the (relatively) modern world. As far back as the sixteenth-century, we have evidence of one man, at least, who determined to explore the Alps, for the sake of knowledge alone. Conrad Gessner was a Swiss naturalist who voluntarily gave himself the task of devoting whatever time he could spare to climbing some of the great alpine peaks, which were little explored at the time. As evidence of this fact, even the word alp is something of a misnomer, being derived from the word the peasants used to describe the high pastures and not the peaks themselves, which they had no use for. While it should be added that Gessner made his journeys in the summer, it was an undertaking notable for the essential desire in people to increase their knowledge of the world. Anyway, snow is present all year round here in the Alps, in the form of the glacier, and it is the glacier that shows that while sometimes people venture into the snow, at other times the snow comes to see them.

We saw earlier how the Little Ice Age had plunged Europe into a long cold period, lasting several hundred years. During this time, there was an ebb and flow of the cold, and with it an increase and decrease in the size of the Alpine glaciers, including one near Chamonix, the Bossons glacier.

At various points during the Little Ice Age the glacier grew in both depth and length, descending towards the village of Chamonix. It caused great problems for the local people, having destroyed barns, outlying houses and even threatened small villages. In 1690, they'd had enough, and paid the Bishop of Geneva, Jean d'Arenthon, to come and exorcise it. Seracs (large house-sized blocks of ice) were breaking away from the tongue of the glacier and destroying crops, as well as causing waters that flooded vast areas. The exorcism was not an unfamiliar tactic; in fact, throughout the century, at each encroachment of the ice, local bishops were called upon to perform this task. On one such occasion in 1644, the warrant officer of Geneva, Charles de Sales, led a procession of 300 people to the foot of the ice. Such processions were still occurring as late as 1818.

It is not an exaggeration to say that the local people at the time saw the glacier as a living thing, or at least an inanimate object inhabited by demons. Inanimate is the wrong word, however, for whatever else a glacier may be, it is not inanimate. It is always moving, shrinking, growing, just at extremely slow speeds. A sleeping giant, the power of the glacier is implicit in the landscape not just here, but across Europe, glacial valleys from the last Ice Age proper being the result.

That the glacier was seen by some as a devil is just one

brief example of an innate instinct we seem to have and one that I find deeply beautiful; the desire to make stories, stories about everything and anything, and it is myths and legends about snow that I'll turn to next.

It is still snowing, but we have had good news. The road has been declared safe to open for an hour or two each day while work continues in the meantime trying to prevent the problem from recurring. The *Mairie* might want to consider the exorcisms of earlier times; apparently the glacier did recede for some time after Jean d'Arenthon's intervention in 1690. So perhaps it's possible to fix this weak spot on the road. If it is, however, and whether it's with spells or concrete, it will only be for the short term, that is to say, the lifespan of mankind. Nature's timescales are somewhat longer than our own, even than that of our species. Nature has a way of doing what she wants to, in the end, and it only needs a little land and snowslide to block a road to remind us of that.

Snow Queens

S now is a rare thing.

To date, less than 600 people in the history of the world have travelled into space, and been able to see the Earth from outside. It's a feeling the rest of us can snatch a glimpse of thanks to the live webcam feed from the International Space Station. Ninety minutes is all it takes to complete one orbit of our planet, and it's a dangerously mesmerising hour and a half to spend, virtually, gazing at our home through the lenses of the Earth Viewing Experiment. One thing that's noticeable, given an absence of clouds on your flight, is how rare snow and ice are on our planet. It's not surprising therefore, that myths and legends about snow are also correspondingly rare.

Consider this: *The Golden Bough,* by Sir James Frazer, is an unequalled study of mythology and comparative religion. Mention it to anthropologists now and you will most likely be met with a sneer; times have changed and Frazer has been deeply criticised for armchair research amongst other terrible crimes. Yet it remains a remarkable achievement; the third edition, published between 1906 and 1915, runs to twelve volumes, and all because Frazer wanted an answer to one simple, if rather odd, question: why were the pre-Roman priest-kings at Nemi ritually murdered by their successors? The answer is a matter of rebirth and reincarnation that Frazer came to believe is common to almost all the world's religions, and it took him twelve books to establish that, taking in stories, mythologies and religious and magical practices from all around the world.

The index and bibliography alone comprise a thirteenth volume. Look up the word 'rain' in that index and you will find around a hundred entries. For 'sun' there are even more. And 'snow'? In all the twelve volumes, snow merits just one mention: a story from Northern Pakistan about an evil king whose soul resides in the snow. That's the rarity of snow. There *are* myths and legends about the snow, of course; they're just a little harder to find than other kinds of story, and in this chapter we will go in search of one or two.

Before we start, however, it's obvious that old stories, by which I mean myths and legends, will to a large degree be influenced by the world which created them and into which they were born. Furthermore, I think they can be considered living things, with their own genealogy and DNA. They are passed down from generation to generation, changing and mutating as they do. Recent research using mathematical modelling suggests that fairy tales, such as the ones found in Brothers Grimm's *Household Tales* for example, are not just a few hundred years old as has typically been thought, but may be thousands of years older. Wilhelm Grimm thought as much himself, believing that they may have originated long ago in places ranging from Scandinavia to Asia, but even he may have been surprised to learn that some of these tales are possibly as much as 6,000 years old.

Maybe this shouldn't surprise; we still tell our young school children the story of Theseus and the Minotaur, for example, a story that is *at least* 2,500 years old and possibly much more. Nevertheless, what this research demonstrates is worth considering for a moment. It means, for one thing, that the languages we think our fairy tales were written in – English, German, French, Italian – didn't even exist when the stories were born. Instead they would have been told in that theoretical ancestor of all these tongues: Proto Indo-European. Moreover, if the stories are 6,000 years old, they wouldn't have been *written* at all, since writing as we know it only starts to develop around that time. Instead, they would have been passed on orally, and I believe it's a fair assumption, given the way we humans are, that oral story telling began as soon as we had things to talk about and language to do it with, in some unimaginably distant past.

Old stories such as that of Theseus have survived for thousands of years, living inside us and evolving all the while. I use the term *evolving* deliberately; I think stories are susceptible to Darwinian selection; the good ones survive, the weak ones die out. What do I mean by 'good', and how do they survive? The 'good' stories are the ones that mean something to us, that resonate with us, that continue to show us the world and interpret it for us, even if they happen to be several thousand years old. In this way, they live inside us, and so get passed on to our children, while stories that please us no longer die out. Sigmund Freud's view of popular fiction was that it was the wish fulfilment of its authors; authors enact their desires through their characters and their characters' actions. Freud took this further and saw myths and legends as the wish fulfilment of whole peoples, whole nations. This makes a lot of sense; we keep the ones that mean something to us as a society; the rest are forgotten history.

Heading into a cave of sorts, the labyrinth, as Theseus did, to defeat a beast or monster of some kind: this must be one of the most ancient stories of all. Could it be that his legend is in fact a relatively *modern* version of an indescribably old story? Our ancestors must have sat around the camp fire and told a tale just like this; an account of one of their kind, a hero, who had ventured into the darkness and emerged triumphant, having slain a cave bear or mountain lion. That we are still telling the story now is because we still find it to be a 'good' one, it still means something to us, even if we now see our monsters and caves as metaphors.

Myths, legends, folk and fairy tales have been a big part of my life. I did not leave them behind with childhood; actually, I don't think any of us really do. They remain part of our worldview; we all know who we're speaking of if we mention Snow White or Cinderella, we use references to fairy tales in every day speech ('you shall go to the ball', 'he's a real Prince Charming', 'a trail of bread crumbs') and more importantly, they inhabit the way we see the world. They are also born from the world in which we live; that we do not have so many legends about snow (in England there are very few) is simply because we do not live for the most part in a snow-bound world.

But they do exist. One English story is a trope found across many countries: that of 'snow in summer'. St Bees in Cumberland is said to be the site of a miraculous intervention by God; the longest version of this story is given by Edmund Sandford in his *A cursory relation of all the antiquities and familyes in Cumberland* c. 1675. A pious abbess and her nuns were driven by a storm into Whitehaven where their ship ran aground. Taking refuge in Egremont Castle, the Lady of the house asked her husband to help them; the Lord decided to make merry at their expense:

> *And the Lord laughed at the Ladyes : And said he*
> *wolde give them as much Land as Snow fell upon the*
> *next morning bein midsumerday: And on the morrow*
> *looked Out at the Castle window to the seasid 2 miles*
> *from Egremont all was white with snow for three miles*
> *Together : And thereupon builded this St. Bees Abbie.*

A very similar story is told of the founding of Wroxall Abbey in Warwickshire. Perhaps, although we're told the snow fell on Midsummer Day, such stories grew from actual freak weather events in the Little Ice Age. I like to think so. There's even an old name for late wintry events such as this: when snow falls in the spring, it's called a blackthorn winter from the bush that's in blossom at that time. (Incidentally, those saints whose saint days fall during the blackthorn winter are known as ice saints, or frost saints. They are Saints Mamertus, Pancras, Servatius and Boniface.)

It pleases me a lot to think that many of our myths and legends may have derived from real people and real events, albeit exaggerated and twisted by the evolution that stories naturally undergo. It also makes more sense than thinking that storytellers just 'made them up', given that stories have to mean something to us in order to survive, as I argue above. For example, about ten years ago I made a trip to the Maramureş, that most traditional region of Transylvania. I was speaking to my guide about such things, old stories and their history, and he told me about The Winter King, a legend of the Middle Ages, long before Romania itself existed. That era saw prolonged invasions from the south by Ottoman armies, and my guide told me the story of how one invasion faltered and failed due to the harsh winter. The Turks were routed and many perished, lost in the snowy forests of the Carpathians. Before long, my guide claimed, the story became a legend; the forest itself became the Winter King, a supernatural force that would protect the people from further Turkish attack.

I've never been able to find a single other reference to this story. I don't know if my guide made it up himself, or whether it's a little-known story that had been passed down to him, but either way, I don't really mind, and either way, there's a name for this process of history becoming legend; it's called euhemerism, after Euhemerus, the Greek mythographer.

Another asserted case of euhemerism concerns that

of Odin and the Aesir, the gods of the Norsemen. The thirteenth-century account of their origins, given by Snorri Sturluson, suggests that Odin and his kin were real people who came from the area around the Black Sea, and slowly emigrated north and west, finally arriving in Scandinavia, where they so impressed the local people that they later came to be revered as gods. This story was put to the test by the Norwegian adventurer, Thor Heyerdahl, himself a near-legendary figure by the time he undertook a series of archaeological investigations in Azerbaijan, towards the end of the millennium. He died before he could find any truly conclusive evidence, sufficient to prove the matter to the archaeological community that is, though there are some tantalising rock carvings and written sources that seem to hint that Sturluson's story was, just perhaps, true.

It's a simple, appealing and powerful notion; we worship the people who came before us, as gods. How much truth there is in that statement remains unclear. One thing is for sure, however, that legends of snow are much more common amongst the people of the north; as Robert Graves had it: 'the quarter from which the wind brings snow; only dead suns are to be found in the cold polar north.'

Ice is central to the Norse account of Creation, for example, again told by Snorri Sturluson. Ymir, the primeval being, was suckled by a cow called Auðhumla. As she nourished Ymir, she also licked salty blocks of ice, from which three days later emerged Búri, the first god, from whom all others descended.

Snow is used as simile and metaphor in the tales of the Norse. In *Egil's Saga*, one of the *Sagas of the Icelanders*, the headwind is described as having a chisel of snow. Elsewhere, in the verses of the *Poetic Edda*, silver is described as 'snow-shining'. In the same writing, in the verses known as *All-Wise's Sayings*, Thor comes across a dwarf who wishes to marry his daughter, a girl described as 'snow-white'. The name Thor itself merits a mention. Thor is the personification of thunder, that we know, but his name is related to that of another god: Thorri, the personification of frost and of midwinter. Modern thinking holds that Thorri is a diminutive of Thor, and that Thorri became associated with midwinter since the yearly sacrifice, or *blót*, in Thor's honour was held at that time of the year; the *Thorrablót*.

Wherever there is plentiful snow, stories of it abound. Northern tribes of Native America possess many such tales, for example the Lenape legend of Snow Boy, a frosty figure who is taunted by other children (in particular, they poke his frozen penis). In return, he sucks their fingers and toes, leaving them with frostbite, a situation that continues until he's tricked into floating away downriver, a common enough remedy for problem elements in folk and fairy tales.

The snow child is a fairy tale found in many variants, one recorded as type ATU 1362 according to the Aarne-Thompson-Uther folk tale classification system. There's an English version of this story, in which a woman falls in deep snow on her way home. When she returns to her husband, she informs him that she is with child, and gives birth to a boy 'as faire and whyte as Snow itselfe'. Because the wife is 'worth the looking on', the husband sets aside all sorts of misgivings about the boy's paternity for the time being. The boy lives with them happily enough until one day when the husband sees his chance and takes his icy son for a walk in the warm sun; he melts 'all into water' and that's the end of Snow Boy. And the marriage too, in the retelling by Sir Thomas North, originally published around 1570. Although later sources disputed the claim, North himself held that the story originated in India, with the legendary philosopher Bidpai, a figure, who like Aesop, may or may not have been real. Given the research mentioned above on the origin and age of some tales, like Wilhelm Grimm, North was perhaps closer to the truth than he was given credit for.

Another common trope is ATU 703, The Snow Maiden. This is a story predominantly found in Russia, and one with many versions even there. Sometimes entitled *Snegurochka*, it exists in ballet, opera, film and ballad versions, and was included by Arthur Ransome in his collection *Old Peter's Russian Tales,* published in 1916, as *The Little Daughter of the Snow.* This was one of my favourite books as a child. Having read every *Swallows and Amazons* book there was, I wanted to read more by Ransome, and so came across Old Peter, which I had on almost permanent loan from the library. At the time, I knew nothing of its genesis; Ransome, in an effort to literally escape his first, disastrous marriage, decided to spend some time abroad. He chose Russia because it was one of the few places that required a passport, and knowing that his wife, Ivy, would have difficulty getting one, he fled to the east. He seems to have fallen in love with Russia, with its people, its language and it stories, and having learnt the language, decided to write a book of Russian fairy tales for British children to enjoy. Even now, the skill of it is delightful. Knowing that some concepts would be alien to children back home, he frames the tales within that of a grandfather

(Old Peter) telling the stories to his grandchildren (Vanya and Maroosia) by the fireside of their little hut in the vast Russian forest. This simple conceit provides some of the most enchanting passages in the collection; for example, at the start of *The Little Daughter of the Snow* Old Peter describes how the old couple of the story are very unhappy because they have no children.

'Think of that,' says Old Peter. Though he adds, 'Some would say they were better off without them.'

'Would you say that, grandfather?' asks Maroosia, to which her grandfather replies wryly, 'You are a stupid little pigeon.'

Old Peter tells his grandchildren the story; how the old couple fashion a child from the deep snows, a girl who is to be their very own daughter, though she tells them she must never go inside their hut, because of the warmth there. She also warns them that if ever the time comes when they do not love her, she will melt away again. Sure enough, by the end of the story, when it seems the couple are more concerned about an escaped hen than their daughter, she runs into the house and melts away by the fire.

This is in contrast to some versions of the tale, where it is love itself that makes the snow maiden melt. For example, in the play by Aleksandr Ostrovsky, with incidental music by Tchaikovsky, it is her own love for a shepherd that causes the snow maiden's heart to thaw and her body to vanish.

With frozen hearts, we now we begin to approach the most familiar female face of winter; that of the Snow Queen. It's true that snow and ice are personified in many male forms; in addition to Thorri and the Lenape Snow Boy, we might mention Snegurochka's grandfather (according to modern variants), a Father Christmas figure known in Russia as Ded Moroz, meaning Old Man Frost or Grandfather Frost. There's also of course Jack Frost, who leaves us with frozen windows and nipped fingertips and noses, a figure of such widespread fame and location that it's very hard to be sure of his origins. On that basis, one might suspect he is a very old personification indeed, yet the psychologist G. Stanley Hall, writing in his mammoth two-volume work on adolescence, reckoned that 'Jack Frost comes as near or nearer than anything else to an independent modern creation of the child mind.' Either way, young or old, Hall felt that Jack Frost filled 'a real need of the childish soul... vastly more plastic and less conventionalised than Santa Claus.'

Despite the existence of male apparitions of frost, snow and ice, I think they pale (if that's the right adjective) in comparison with that of the Snow Queen. In Hans Christian Andersen's *Snedronningen*, *The Snow Queen*, young Kay has been bewitched by splinters of an evil troll's mirror, and the cold of the queen herself. It is Kay's friend Gerda who eventually finds him at the Snow Queen's palace, in the middle of a frozen lake. Gerda's love saves him: her warm tears melt Kay's heart, burning away the splinter from it. He bursts into tears and this dislodges the second splinter in his eye. They return home, safe, though it's noticeable that the Snow Queen, though defeated in her possession of Kay, is not destroyed. The possibility that she will return always remains, just as it does with snow itself.

Andersen's *The Snow Queen* is the clear inspiration for Jadis, C.S. Lewis's 'White Witch' – the model of the abduction and imprisonment of a young boy is the same, both are sexually alluring to a degree. Kay is seduced by two kisses from the Snow Queen, Edmund by the forbidden fruit of Turkish Delight. Both are powerful and beautiful women, and, above all, they are cold. Very cold.

We should not be too hard on Lewis for displaying no great originality in his conception of Jadis; both the power and the weakness of working with archetypes is their very familiarity. It's thought that Lewis was also influenced by other malign women; H. Rider Haggard's *She* and George Macdonald's *Lilith*, for example, cold women all. Today, to call a woman an ice queen is to make a very explicit statement of their lack of empathy and frozen emotions. There is also the intimation of a deadly kind of beauty. This connection between warmth and love, and snow and heartlessness, spans all countries; and perhaps finds its most chilling form in the Japanese figure of Yuki-Onna.

As with all tales, there are many variants, yet the common thread throughout is that of a tall, mysterious and beautiful woman who appears in the snow. Sometimes she is naked, sometimes dressed but seemingly without feet (Japanese ghosts are frequently described as having this trait). In the version recorded by Lafcadio Hearn, one of the West's earliest transmitters of Japanese culture, Yuki-Onna appears to two woodcutters, one an old man and the other young. Sheltering from a snowstorm, the two men are visited by the alluring figure of a snow-white woman. She breathes on the old man and freezes him to death, but spares the young man when she sees that he is handsome, on the condition that he must never speak of what has happened. Surviving his ordeal, he grows a little older, and one day meets and falls in love with a beautiful young girl. They have children, and all is well until one evening, while his wife is sewing by the firelight, the young man remarks that she reminds him of a strange person he once met . . . Thereupon, all is undone. He has broken his promise never to tell. His impostor wife reveals herself as Yuki-Onna, who only spares his life this time for the children's sake. She disappears into the night, never to be seen again.

Yuki-Onna is a massively popular figure in modern Japanese culture, star no longer of old folk tales but nowadays of Manga comics, video games, and films.

Time moves on. Words pile up, one by one, like snow flakes, each seemingly adding almost nothing, and yet gradually the valley fills with a drift. If we do not move on from this world of myth and legend, we may find ourselves like certain characters sometimes found in folklore; those waking in snow were said to be the victims of fairy enchantment. Having been tricked into thinking they were being entertained, wined and dined in a sumptuous house, they wake in the snow in the morning, blinking, rubbing their eyes, and know they were fooled by elves and their kin.

This would be a good place to return to our cousins from the north. I have always felt that we neglect the culture of these cold lands. The preference for Greek and Roman myth is one we owe perhaps to Victorian education, and a lasting bias about what makes empires great. While I would not wish to dispose of the Classics, I feel there's a genuine place for northern mythology in our classrooms; stories of people very close to home, who invaded and settled in the British Isles, giving us much of our language, place and family names, and cultural heritage.

The Exeter Book is a collection of Anglo-Saxon poetry and riddles, over 1,000 years old. The poems, such as *The Seafarer*, are magnificent, always real and vibrant; the riddles are by turns strange, obscene, clever and cunning. The meaning of some of these is easy enough to guess, while others remain the subject of dispute amongst scholars. Perhaps the difference is merely that some of the riddles are 'true riddles' while some of them are 'neck riddles' – the former it being possible to work out, the latter having an answer known only to the setter, and thus which might be used to save one's neck, just as Bilbo Baggins does in his battle of wits against Gollum.

Amongst the complete verses of the Exeter Book there are a number of fragments, one of which is a one-line riddle, usually denoted as number 69.

The one-line riddle reads like this:

> *Wundor wearð on wege   wæter wearð to bane*

As always, it sounds better in the original, which in this case is Old English alliterative verse, but for once it can be very decently translated, like this:

> *On the way a wonder; water became bone.*

The answer to this riddle captures the essence of the final aspect of the snowflake: opposites and transformations.

Transformations

Even before we open our eyes we know that snow has
come. The sound of the valley has changed. We sleep
with the door to the balcony open a fraction, no matter what
the weather; for most of the year we hear the unending
sound of the waterfall down below us, in the hidden heart of
the valley's forest. At some point each winter, however, there
comes that first morning when the singing of the cascade has
gone – if enough snow falls as we sleep, it blankets the noise
of the rushing water sufficiently for it to appear to us to have
stopped altogether, overnight.

Still before we open our eyes, there is enough bright
reflected light pouring into our room to discern the

difference from a normal day, even through closed eyelids. Seeking confirmation, we sit up in bed, and look. We gaze at it, we drink it.

As of yesterday, the season's snow had finally melted. Below the house, in the meadow, spring could be seen arriving in an energetic rush, desperate to make up for lost time. Yesterday, a warm spring rain washed the last fugitive piles of this season's snow from even the shadiest corners of the terrace.

Yet after several days of fiercely warm days during which the meadow flowers opened and leaves burst forth on the sycamore trees, softening the predominant evergreens, winter made one last appearance. Yesterday, a sudden cold snap and bitter precipitation encrusted the mountain once more in a complete sheet of white, and put a layer of snow on our doorstep that still remains.

By rights, we should have expected a last visitation of snowfall and even ice this morning, for these are the days known locally as *les saints de glace*, belonging to those frost saints we met earlier. Traditionally it's to be expected that the dying winter takes one last nip at you now, catches you unawares, sends you sliding. It is the second week of May and the frost saints are right on time. Snow has fallen, and the world has changed.

As Robert Graves wrote in the poem that opens this book, there is something mysterious, something enigmatic about the way that snow arrives in the night:

> *She, then, like snow in a dark night,*
> *Fell secretly.*

Following on from thoughts of snow queens, we see Graves using snow as simile for a female lover in this beautiful poem. He continues, capturing the sense of being blinded by a lover, by the snow.

> *And the world waked*
> *With dazzling of the drowsy eye,*
> *So that some muttered 'Too much light',*
> *And drew the curtains close.*

It's too much, this brightness. Like the snow-blindness that threatened the eyes of explorers like Scott and Amundsen; this lover is just too much to look upon directly. And yet, she is not all snow maiden, not all ice queen:

> *Like snow, warmer than fingers feared,*
> *And to soil friendly;*
> *Holding the histories of the night*
> *In yet unmelted tracks.*

Despite her potentially fatal nature, her life-threatening cold, there is warmth in her nevertheless, and it is her very cold which preserves, in *unmelted tracks*, the memory of those warm fingers. . .

In this short poem, Graves captures a wealth of different things. Such is the genius of great poetry, and it is a poem which ably captures the themes of this final chapter as I'll try to collect all these disparate thoughts about snow, and understand what it is about snow that attracts me, and many others, so much. At the close of *Like Snow*, Graves is alluding to one of snow's most perplexing characteristics; it is full of opposites.

Snow is, of course, cold. It begins to melt as soon as the temperature rises above freezing. It is nothing more than crystals of ice after all, and the freezing/melting point of water is zero degrees Celsius. Yet even here the contrariness of snow begins. Snow can survive a long time in the right circumstances, even in direct sunlight, even in above zero air temperatures. This was something I learned as a boy, one winter, when I was eight. The snow was the 'right kind'; perfect for snowballs and snowmen, and one day I built a tiny igloo in the front garden, just big enough for me to sit upright inside. The igloo wasn't built out of carved blocks of ice as an Inuk would make one; I packed the snow into a curving concave hemispherical structure, handful by handful, with walls just a few inches thick, and amused myself by sitting inside and watching the goings-on in the house from my distant polar lair.

I tired of that pretty quickly, and the snow in the garden melted after a couple of days, but my igloo of hard-packed snow lasted for over two weeks. Though it dwindled a little further each day, I was amazed by how long any of it lasted at all in the sunshine and warmth that followed.

The mechanism behind this survival technique of snow is down to a thing called albedo. From the Latin, *albus*, meaning white, albedo is the ratio of light that is reflected from a surface. A surface that perfectly absorbed all light would have an albedo of 0, one that perfectly reflects all light would score a 1. In the real world, everything exists between these theoretical extremes, with dark colours and blacks appearing closer to 0 and whites at the other end of the scale. The ocean, for example, typically has a very low albedo, less than 0.1, meaning that less than 10 per cent of the light that falls on it is reflected and the rest of the energy is absorbed, warming the water. Sea ice on the other hand, has a high albedo, of up to 0.7, meaning that 70 per cent of light that hits it is reflected. Snow scores even higher, with an albedo of around 0.9. What this means is that snow is very good at insulating itself; it absorbs less light energy than a dark substance and therefore maintains its cold, delaying melt.

Snow covering sea ice is therefore a very good combination in self-preservation, and yet, eventually, as the snow melts and pools on top of the ice, the albedo starts to drop very rapidly, from 0.9 down to around 0.15, thus accelerating the melting process. Some of the debates around climate change concern such tipping points, as the stasis-maintaining capacity of the ice caps suddenly is removed from the planet. To return briefly to this discussion; it has been proposed that some of the extremes of the Little Ice Age were due to man's influence, even as far back as the Middle Ages. It's been suggested that large-scale deforestation during this time increased the albedo in snowy regions across Europe, increasing the length of time that snow survived on the ground, maintaining lower temperatures for longer in the process.

This theory is unproven. What's obvious, albeit strange, is that snow is a good insulator. That's something else I discovered in my boy-sized igloo. I was warm in there, really warm. Suddenly, the scarf and gloves that hadn't really been enough to stop me from getting cold in the snowy garden were too much. In order to stay comfortable, I had to take my coat off.

Later that year, the idea that snow was warm was reinforced when my parents gave me a one-person tent for my birthday. It was the second week of April, and I decided I wanted to sleep that night in my new tent, which was a modern looking thing for the time, shaped like a wedge of cheese. That night it snowed, quite heavily. Worried, my mother trudged outside to drag me back into the house, convinced I would be freezing. She could see the shape of my body as the snow had flattened the skin of the tent over and around my sleeping bag – and yet I wasn't even aware that anything was wrong – I had remained asleep, dry and warm.

Here's a considerably more extreme example: in mid-February 2012, north of Umeå, Sweden, on one of the endless snow roads through the forest, a man was found in his car. He was still alive, but barely, having been trapped in the car since December 19, surviving sixty days without food. He'd endured by eating a little snow from time to time, snow that had otherwise kept him warm, being packed all around the vehicle. It was the snow that had trapped him, yet it was the snow that kept him alive.

As Graves hints at in his poem, snow is *to soil friendly*. In cold climates, a blanket of snow on the ground keeps the soil temperature above what it would be otherwise, helping to prevent permafrost, allowing life of all kinds to hibernate, sitting out the winter until the melting snow signals that it is time to emerge once more. The Alpine Springtime is glorious; pastures burst into floral life; gentian plants thrust their stalks up into the air across hillsides riddled with the burrows of animals such as the marmot, who you can see standing at the mouths of their tunnels, blinking in the sunlight, noses twitching in the warming air. All this life held in hibernation, just underground, just under the snow, just under the rush and roar of the skiers carving their way down the mountain a few weeks before.

The stranded motorist in Sweden was extremely lucky, but often, even with all our sophistication and technology, people stranded in snow are not so fortunate, as we saw earlier, and here waits another of snow's most striking pairs of opposites.

It is hard to conceive of anything in the natural world that we can perceive with unaided vision that is lighter than a single flake of snow. The weight of a snowflake can vary enormously depending on its size, water content and so on, but is typically in the region of 0.001 to 0.003 grams. Yet, en masse, snow very quickly starts to weigh an awful lot, as the drooping branches on the firs outside my window testify, and it's no accident that the roofs of *chalet d'alpage* like ours are built to withstand enormous stresses. A local woman in her eighties told us about the heavy winters in the valley when she was a girl. She remembered how she and her family would sometimes climb into their chalet through an upstairs window, five metres off the ground, how the snow sat on the roof, two metres deep, a weight of many tonnes. Yet these old places stand, for the most part, unless they are wiped away by the power of the avalanche, as our friends' house up the valley once was, long ago.

The power of the avalanche is terrifying. This is proving to be a strange winter, a very mild one, but not therefore without dangers from slides of snow. There have been several deaths already this year, the most recent one last week as five Czech skiers were killed. The avalanche was almost two kilometres wide, and five metres high; there is no out-running a thing like that.

Many slides of snow are small sloughs of dry powder, causing no damage, posing no danger to life. Avalanches are their deadlier cousins; their size and power arising usually when a fresh fall of snow puts too much stress on the existing snow pack. Even then, it should be noted that this only prepares the right conditions for the avalanche; it's estimated that 90 per cent of avalanches are caused by human action – it is the victims themselves who trigger a slide of snow, which within a few seconds can reach speeds well over a 100 kilometres per hour, obliterating everything in its path. When it finally comes to rest in what's called the run-out zone, the snow settles like concrete, entombing anything it's caught. The survival figures for those buried by an avalanche are surprisingly good, but only to start with. 93 per cent of victims survive if dug out in the first fifteen minutes, however, the numbers drop quickly. After forty-five minutes, less than 20 per cent of people survive, and beyond that the figures make even grimmer reading.

Snow ranks amongst the greatest forces in the natural world; either in the form of the rapid catastrophe of an avalanche, or in the form of snow's offspring, the glacier, with the power to carve mountains and rearrange the face of the Earth. Such forces are unimaginably powerful, and occur over almost incomprehensibly long periods of time. Nevertheless they are the result of the humble snowflake, tiny and almost weightless. Miniscule, intricately beautiful too, and yet which can be deadly. For many, there is little more alluring than the sight of a snow-capped mountain on a clear blue day; the pull to be out in the snow, in nature, can be overwhelmingly strong, and yet the misery of a white out, the very real danger presented by attempting to travel in a blizzard, show us how horrific snow can be too.

Warm but cold, beautiful but terrifying. Minute but enormously powerful. Such opposites, such contradictions, are what snow is all about.

But are they contradictions? I don't see it that way. Snow has opposites, yes, but these opposites are merely aspects generated by the central meaning of this mysterious entity. This meaning is the heart of snow, and it's the meaning given by the solution to that Anglo-Saxon riddle at the end of the last chapter.

On the way a wonder, water became bone.

The accepted solution being that while riding the waves in a ship, the sea has frozen, turned to ice, white and hard as bone. This is snow's essence: transformation.

It's the transformation that occurs on that morning each winter when the first snow arrives. It's what draws us so strongly when, as Llewellyn Powys wrote: 'we hear that it is snowing, to cross our fire-lit rooms and glance between the curtains out of the window.' The deep, deep silence that has moved in is the creation of the new landscape around us. It's hard not to feel that special sense of otherness that snow produces. Given a heavy fall of snow, in no time at all the world around us appears to change its very essence. Slipping on coat and gloves, we move out into the frozen world in a kind of waking dream – as in a dream, the familiar has become unfamiliar. We look at sights we see every day, our house, the trees around it, and we wonder at it all. Even sound works differently, something we maybe do not consciously notice and which further enhances this sense of the unreal. The air feels different in our noses, in our mouths; our breath vents in steamy clouds; any bare skin tingles. All in all, it's a strange experience, one that starts us thinking, maybe makes our mind move in a different way to usual. We ponder, we imagine, we fantasise, because the world has changed. As Werner Herzog wrote in *Of Walking in Ice*, 'This is a season which is no longer anything Earthly.'

Is it its ephemerality that creates a notion that snow is strange? That it may leave as soon as it arrived, vanishing before our very eyes? When I see heavy snow, am I remembering those falls of snow from childhood, the ones that really were heavier and more frequent? Is it this sense of longing for the past that erupts into the present moment, flooding me with calm, and excitement, simultaneously? Those days when my brother and I missed school because of the snow, those were such days: of calm, because we did not have to face what waited for us inside the school gates, and excitement, because we were free to explore the transformed world outside our house. Is snow, in some way, a form of nostalgia?

I feel sure that's a part of it, but there may be simpler explanations, perhaps the simplest of all being this: snow is white. There is relatively little in nature that is white, outside the polar circles, of course (and I am not going to venture into a discussion about how many words for 'white' the Inuit have). For most of us, natural whiteness is a rarity. Certain flowers, chalk, the crests of waves, clouds. In the heavens: the moon and stars. All beautiful things. All white. And then there is snow, the whitest of them all, which shines so fiercely on a bright day that it cannot comfortably be gazed upon with the naked eye. Is part of snow's wonder simply that it is white?

White is unusual; it is that colour which is at once not a colour, and all colours combined; opposites again.

White symbolises purity, cleanliness, virginity. It is doctors' coats and not so long ago, hospital wards. It is the Pope's vestments. It is wedding dresses and swans. In fact, both the swan and the Virgin were significant symbols to the alchemists, for they represented *albedo;* that term again, but which to those medieval truth-seekers was a transcendent whitening process in the retort, one of the four major stages of the transmutation. During *albedo*, a stage of illumination, impurities were washed away. It was a swan, too, that William Blake chose to illustrate plate 11 of the epic poem that actually owns the title Jerusalem: *The Emanation of the Giant Albion*. (The lines now commonly referred to as 'Jerusalem' are in fact from the introduction to another of Blake's big works: Milton.) The verses of plate 11 show man's struggle against materialism and orthodoxy; Blake depicts how Los, a fallen, earthly prophet and representative of the imagination, is:

> *Striving with Systems to deliver Individuals from those Systems*

. . .the true meaning of 'Satanic Mills' being the orthodox church. And Blake chooses to symbolise this struggle in the form of a forlorn, weakened swan, swimming in a turgid lake – the noble bird being an emblem of creative forces paralysed by the destruction threatened by the 'spectre of reason'.

White is also the colour of milk, and of semen, both bodily substances that were believed to be purifications of the substance called by Dante *sangue perfetto,* 'perfect' or 'divine blood'. Based on Aristotle's theories, *sangue perfetto* was the seed of human existence, and could be refined by the body into both semen (which creates life) and milk (which nourishes it). In this way, whiteness is the symbol of life, of fertility, of the new. It is the colour of innocence too; the Vestal Virgins wore white, and we might think of the lamb, the unicorn; religious symbols both. A variety of world religions deploy white, as humility and devotion; the pilgrims of Shinto and of Islam, the Brahmins of India all wear white.

But white does not mean the same to everyone the world over; it could be pointed out that in some Asian countries, white is the colour worn during mourning, and is taken to symbolise death, but actually there is a little more to it than that. White is the colour of reincarnation, in which life and death are indivisible. Whiteness therefore symbolises life as much as it does death, just as the two halves of the yin/yang equivalence are inseparable.

For a writer however, white means one thing above all else: the empty page. I only see now how, as a child, the blank canvas of snow provided the perfect vacuum into which to pour a fertile imagination. We tumbled through the snow without a thought as to how and why we were creating worlds from inside, planting them in the 'real' world around us. We made stories, we fought battles, we explored remote wastes.

Then, something appalling happened. We grew up. As a professional writer, it's all too easy to see the empty page as the exact opposite of the imagination. During spells of writer's block, an empty page is the biggest evidence we have that the words are not coming. It is terrifying, stultifying, it is sterility and it is a very painful proof of the nothingness we feel will never depart. At such times, it's important, actually it's *vital*, to remember the child in the snow; far from seeing the white page as a threat, we need to remember that it was that very emptiness which was exciting; that emptiness is the indication that all things are possible, that everything and anything in the world (or beyond) can be written about; the blank page waits for us to fill it with something new, unknown and unexpected, and from the twenty-six letters on our keyboard, make magic. (Magic? To frighten people, to make them laugh, to move them, to make them think; surely it's alchemy to do all that with just the twenty-six letters of the alphabet? And, as someone once rightly pointed out to me, some punctuation.)

Opposites again. And one of the most significant ones to me as I fill a few more lines of this page with small black squiggles. Is this why I have come to believe I love snow even more than I did as a child? Or is it something else that has escaped me entirely, something that will only occur to me a year or two after this book has been printed? That happens sometimes, with books. They are, after all, only a snapshot of what is in your mind at a given time.

★

Too many questions. I think it's all the things above, and more. Once again, it's worth noting the opposites I used to feel when confronted by snow, and still feel; simultaneous calm and excitement, and worth appreciating that not only can we hold opposites inside ourselves, we can hold them both at the same time. And that's a sure-fire way to feel strange.

Maybe it's simply the comparative rarity of snow that makes us feel this strangeness. Perhaps, but as we saw in the chapter on mythology, even those cultures that live permanently with snow still appear to revere and mythologise its nature; its coolness, its deadly beauty. As we saw, there are fewer myths and legends about snow than about many other natural phenomena, but if snow is the most visible emanation of a wider phenomenon – winter – then folklore does indeed have a lot to say about the subject.

All ancient cultures, and if we delve a little, most modern ones, still exhibit their reverence for critical points in the cycle of the year: midsummer and midwinter. That we still use the word Yule (the old Norse midwinter) to signify Christmas, that we bring greenery into our houses in the form of fir trees, that we burn a Yule log (or eat a chocolate one) or light candles; that we exchange presents, that we eat the finest food we can lay our hands on; all these things are signs of an ancient and pagan reverence for midwinter. (As Sir James Frazer was finally able to ascertain in *The Golden Bough*, the core to all ancient religions was the notion of rebirth, very often symbolised by whiteness.) It is here, at the shortest day of the year, midwinter day, that the world seems to have died, and from where it begins to grow again, as the days literally grow longer, the sun warmer, plants and animals to burst forth with renewed life, as if reincarnated, as if death has been transformed into life.

For those of us where snow is a seasonal thing, albeit either common or rare, the transformation from one state to the other is the very core of snow, as well as, I think, the reason for the sense of otherness. Freud's essay on strangeness, on fear and dread, *Das Unheimliche,* is usually translated as *The Uncanny.* In it, he developed the idea of the uncanny as discussed some thirteen years earlier by Ernst Jentsch. It's one of those cases yet again where no suitable direct translations correspond between the two languages. In German, the pair of words *heimlich/unheimlich* are a clear pair of antonyms, a pair which does not exist any longer in English. Freud's theory of the uncanny revolves around this pair of opposites. If *heimlich* means something like homely/welcoming/familiar, then *unheimlich* means disturbing/unfamiliar/strange. Why this is important is because it is that which is *nearly* normal which has the power to disturb us the most. This is especially true when the thing in question is a simulacrum of ourselves, which explains why we find dolls, automata, clowns and such things creepy. The idea has given rise to the term *uncanny valley*, used in robotics and elsewhere, to describe our growing sense of unease as an artificial figure looks and behaves *more* like a human being. When we look upon a familiar landscape, but one transformed by snowfall, it is the near normality and yet utter difference that reaches right into our senses. We perhaps struggle to understand this transformation from the mundane to the magical; at one and the same time we find it beautiful and captivating, as well as strange and unsettling.

As this drift of words comes to an end, I want to return briefly to a snowbound mountainside in the Swiss Alps. We left Hans Castorp, the hero of Thomas Mann's *The Magic Mountain*, lying in the lee of a hay barn, trying to endure the blizzard in which he has been caught. Hans Castorp survives his encounter with the snow. Waking from his stupor, he manages to make it back to the safety of the sanatorium, chastened, perhaps, but with something else about him too, something else *inside* him.

As he lies in the snow, close to death, he witnesses visions both beautiful and then terrible. His visions are of heaven and hell: the bucolic scenes by the shore, the nightmare of the cannibalism; we understand that he witnesses archetypal extremes of life and death, a dualism which transcends the combative dialectics of his two philosophical mentors, Settembrini and Naphta, back at the sanatorium.

Hermann Wiegand, who wrote the first long study of Mann's epic novel, saw it like this: 'hovering there between life and death, Hans Castorp is for a moment elevated to a position that marks the acme of his capacity to span the poles of cosmic experience.'

This vision is brief and fades, Hans Castorp recovers from his faint and makes his way back to the real world, but the experience has changed him. It is no coincidence that Mann chooses the catalyst of this change to be a bed of snow, for it is here, amidst the snow, that Hans Castorp is free to find his true nature for himself.

Snow is transformative. It changes the world around us, both in the short and the long term. Overnight, it repaints the landscape in white, inevitably bringing to mind concepts like purity and clarity of thought. Over the course of eons, it can remake the Earth itself.

But snow is more than that. It is full of dualisms; as I look out at the cold landscape beyond the window now, what do I see? Do I see emptiness, isolation, nothingness? Do I see death? Those things are surely there, but so are other things; good things, like peace, silence and ineffable beauty. To live in snow, even for a day, is to be reminded however unconsciously that the world is not always as we see it; that it can change in a moment, for better or worse, and that, most confusingly of all, it is composed simultaneously of opposites, just like the snow. Snow, which is both light and heavy, cold and warm, beautiful and terrifying.

Snow is more than transformative; it is transformation itself.

When my wife and I first met, we were driving around the northern side of Lake Geneva, where the motorway sweeps in and out of tunnels high above the lake. In one particular spot, between two tunnels, there is a short window on the most magnificent view imaginable, high across the water to the Alps on the French side. It was November, it was sunny and the sky blue; the mountains across the lake looked at their utmost divine, and I use a word with religious connotations here with intent.

Just before we were returned to the darkness of the oncoming tunnel, I said one word, that's barely a word at all: 'Wow'. Not so eloquent, for a writer, but Maureen later told me she started to fall in love with me then, simply for feeling the same way she does about nature, and its power. And being speechless before it, as before some kind of god. I have never been one for deities in the heavens, but if I were to worship anything, it would be something as magnificent and celestial as a mountain, or the sun. Or the snow.

We came to the mountains, five years ago, to find solitude; we knew that snow would be a big part of our lives here. We did not know then that I would become ill, and remain ill, from an undiagnosed infection contracted in Asia. But of course, we never know what lies just around the corner. Living in the snow is at times hard, when there seems to be a mountain of snow to be cleared just to leave the house, and the exhaustion that comes with the illness makes all these things that bit harder. It puts more pressure on Maureen, to remain fit and strong enough to do all the work on the days when I can do nothing. Despite that, there is still the sublime beauty of the snow-bound world; the view from my study window alone is enough to take one's breath away, and perhaps do a little more. The power of nature is not just something in our heads: recent research has demonstrated that we have a positive endorphin response to a beautiful view. That's worth thinking about for a moment; our bodies physically respond to the beauty of nature, and they do so with *happiness*, and therefore, health. Hoteliers have known for centuries that this is something you can monetise; it's why we pay more for the rooms in the hotel that face the lake, the sea, the mountain. It's also been proven, having controlled for other factors, that patients on the sunny side of a hospital ward, patients simply with a view from their bed, recover faster than those that do not have the benefit of such things.

As I look out of the window today, this spring morning, snow is losing the fight with the warming year. Maybe one day, in the near or distant future, it will lose the fight forever, and the very last snowflake will melt away to a humble drop of water. Yet until that day, there is always the promise of the snow's return, and when it does return, it brings a message, if only we knew how to read it. For maybe there's more waiting for us in the snow than we first imagine. Waiting for us to understand it, there is the message that life can be transformed at any time, not only for the worse, but also for the better; that nature in itself can heal us, mend us, and that this white rebirth can arrive at any time, perhaps over the course of one dark night. Like snow.

Little Toller **Monographs**

Our monograph series is dedicated to new writing attuned to the natural world and which celebrates the rich variety of the places we live in. We have asked a wide range of the very best writers and artists to choose a particular building, plant, animal, myth, person or landscape, and through this object of their fascination tell us wider stories about the British Isles.

The titles

HERBACEOUS *Paul Evans*
ON SILBURY HILL *Adam Thorpe*
THE ASH TREE *Oliver Rackham*
MERMAIDS *Sophia Kingshill*
BLACK APPLES OF GOWER *Iain Sinclair*
BEYOND THE FELL WALL *Richard Skelton*
HAVERGEY *John Burnside*
SNOW *Marcus Sedgwick*

In preparation

LANDFILL *Tim Dee*
LIMESTONE COUNTRY *Fiona Sampson*
THE FAN DANCE *Horatio Clare*
SEA SOUNDS *Cheryl Tipp*
EAGLE COUNTRY *Seán Lysaght*

A postcard sent to Little Toller will ensure you are put on our mailing list and be amongst the first to discover our latest publications. You can also subscribe online at **littletoller.co.uk** where we publish new writing, short films and much more.

LITTLE TOLLER BOOKS
Lower Dairy, Toller Fratrum, Dorset DT2 0EL